CULTURE AND THE COURSE OF HUMAN EVOLUTION

CULTURE AND THE COURSE OF HUMAN EVOLUTION

GARY TOMLINSON

THE UNIVERSITY OF CHICAGO PRESS
Chicago and London

The University of Chicago Press, Chicago 60637
The University of Chicago Press, Ltd., London

Published 2018
Printed in the United States of America

27 26 25 24 23 22 21 20 19 18   1 2 3 4 5

ISBN-13: 978-0-226-54849-4 (cloth)
ISBN-13: 978-0-226-54852-4 (paper)
ISBN-13: 978-0-226-54866-1 (e-book)
DOI: https://doi.org/10.7208/chicago/9780226548661.001.0001

Library of Congress Cataloging-in-Publication Data
Names: Tomlinson, Gary, author.
Title: Culture and the course of human evolution / Gary Tomlinson.
Description: Chicago : The University of Chicago Press, 2018. |
Includes bibliographical references and index.
Identifiers: LCCN 2017049889 | ISBN 9780226548494 (cloth : alk. paper) |
ISBN 9780226548524 (pbk. : alk. paper) | ISBN 9780226548661 (e-book)
Subjects: LCSH: Human evolution.
Classification: LCC GN281.T66 2018 | DDC 599.93/8—dc23
LC record available at https://lccn.loc.gov/2017049889

♾ This paper meets the requirements of ANSI/NISO Z39.48–1992
(Permanence of Paper).

CONTENTS

PREFACE

This book is aimed at both scientists and humanists, and it contains provocations for each. Its subject matter is the evolutionary emergence of modern *Homo sapiens* over the last 200,000 years, and it advances a model that gives a leading role to the development of culture and human capacities to make it. In itself, this emphasis is unprovocative. The interaction of biology and culture has featured in many scientific accounts of human evolution across the last several decades, and humanists have recently begun to attend to these. Where the provocation may enter in, however, is in my conviction that neither camp has adequately taken the measure of the other.

On the one hand are evolutionary scientists, population geneticists, and paleodemographers who, when they approach culture, usually do so from a quantifying perspective that has to be schematic and blunt. Their model of culture is one that allows its phenomena to be entered as coefficients into mathematical equations in order to track general trends, probabilities, and points of equilibrium. These kinds of models have important uses. They tell us much, and I will exploit them. But they have sheer limitations too, and their builders usually recognize that they cannot explain in nuanced fashion how, in a world populated by several species of encultured hominins (not to mention other animals with minimal cultures), humans came to be the *kind* of cultural animals they are today. Because the models are mathematical, they inevitably form thin descriptions of human attainment and history.

Humanistic scholars—of which I am one—have other things to answer for. Our failing is not our widespread habit of ignoring scientific knowledge in our work. That is fair enough; humanists and scientists produce knowledge with methods and tools that only partially overlap, and cultural interpretation and theory have been able to move very far without recourse to conventional science of any sort. I am not advocating, in other words, that all humanists retool themselves as scientists, or that it is a general humanistic duty to narrow the methodological distance between the humanities and the sciences. Where, however, humanists *do* take on evolutionary issues, it is not enough for us to dabble. By staying in the shallows of evolutionary science, humanists create endeavors of questionable value, wherein a simplified Darwinian paradigm is applied to literature, music, painting, and more—as if evolution could be innocently approached and the complexities of the "modern synthesis" and "extended synthesis" in evolutionary thought bypassed. Whatever the conversations are that need to be struck up with scientists, this kind of work has not helped to put them on a firm footing.

In the middle, between the scientists and humanists, are the Paleolithic archaeologists themselves—meticulous workers in the earth who, with practiced eye, honed skills, and abiding patience, piece together long-dead populations of humans and their lifeways. To linger with them at their sites is to marvel at the richness of the accounts they conjure from traces all but invisible to the uninitiated. They rely on sophisticated science for the analysis of what they unearth, and at the same time they look to the human sciences to help them understand the activities whose traces they read. But exactly in the arena of human culture archaeologists are not well served by either scientists or humanists. Scientists encourage them to make confident statements founded on schematic, quantified modeling, while humanistic influence often leads to a less-than-precise use of terms such as "symbolism," "art," "religion," and "culture" itself—thus encouraging the archaeologists' own brand of dabbling.

This book aims to work a mediation between the sciences and the humanities and to bridge their differences of method, aim, practice, and intellectual habit. Few areas of inquiry call more urgently for such an approach than the question of the evolutionary emergence of modern humanity. The sciences concerned with evolution have changed markedly over the last four decades. These changes stretch conventional analysis to address the behavior of complex systems, the circular causality of feedback networks, and the emergence of features in such systems that are not readily explained by analysis of their component parts. In taking on the implications of all these, evolutionary science has opened new vistas on everything from the origin of life to the rise of consciousness and mind, and humanists now need to take in these vistas in thinking about human evolution. On the other hand, as the mention of mind suggests, inquiry into our origins must engage with all we can understand about characteristic human capacities and behaviors and place these against the backdrop of other animal capacities in order to trace the coalescence of our modernity. Interpreting these things is the broadest charge of the human sciences, and evolutionary inquiry needs to rely on humanistic expertise and knowledge. Conventional science, new evolutionary theory, and humanistic theory and interpretation: these are the pillars that will hold up the bridge; it is paved with archaeologists' findings.

Culture and the Course of Human Evolution is informed by recent trends that engage the attention of an avant-garde of sorts in humanistic and anthropological thought. The biological "turn" in such thought has promoted a number of novel endeavors. It has encouraged humanists to explore in sweeping fashion the place of humans in the world, investigating the cognitive consequences of our immersion in networks of material potentials and constraints—"embodied cognition"—and, following from this, confronting technology, or *technics*, in order to understand consciousness and ideation. It has also led us to conceive of human experience as nested within the experiences of a broad array of species, a perspective that has brought to the fore questions of the relation of human thought, feeling, and affect to those of other organisms. Together these two agendas, the technical and the transspecies, have blurred the borderlines between human experience and both nonhuman life-forms and machinic regimens. Today humanists describe *posthumanisms* and *neomaterialisms*, explore the non- and parahuman, and pursue new fields such as cognitive geography and ecology. All these developments follow on long preparation in the human sciences, and we can discern in them the legacy of work from the 1960s and 1970s such as the "ecology of mind" described by anthropologist Gregory Bateson or the "affordance theory" of psychologist James J. Gibson. But they have taken on a new importance since the turn of the millennium and caught up many humanists in projects that have framed discipline-forging innovations.

What is lacking from most of this new humanist work, however, is a temporal extension into human deep history. Only from such an extension can we hope to explain how a species arose that could shift in fundamental ways the potentials of its own action in the world, or how matter and thought came, through the complex interplay of culture and evolution, into relations that are peculiar to modern humans. The blurring of our differences from other animals and the recognition that we share with them fundamental features of our immersion in the material world call for a historical account of the emergence of our unique position. A considered *post*humanism must address itself to a deep *pre*humanism.

Culture and the Course of Human Evolution aims to sketch such a historical account in terms that take on the specificities granted by both evolutionary modeling and archaeological reconstruction. From this perspective we can explore a place where human activities and capacities, including technology, sociality, and the signs that enable cultural production, merge with those of nonhuman species. Conversely, along the deep-historical axis we can construct plausible narratives for the emergence of the capacities that set modern humans undeniably apart from all other species. The role of a deep-historical vantage in posthumanism is two-sided, both a corrective to anthropocentrism and a nuanced account of how human difference arose.

The thinking behind this book took its start from an earlier one, *A Million Years of Music: The Emergence of Human Modernity* (2015), in which I described the long prehistory of the development of the capacities required for human music and their ultimate coalescence, within the last 100,000 years or so, into modern musical behaviors. To take in this prehistory, the book needed to embrace the million-year span of its title, reaching back long before our species emerged, since (I maintained) only such a long view of pre- and protomusical developments could tell the story of human musicking in a meaningful way. This lengthened historical perspective meant that I was called upon to describe the place of nascent musical capacities in the formation of modern humanity in general, with reference to its major features: language, advancing technology, ritual, and the imagining of immaterial, metaphysical realms. It required me, moreover, to state my views about the general outlines and mechanisms of human evolution across the period, and particularly my position on the relations of biological evolution with the cultures that our ancestors formed and practiced in growing complexity. In other words, the book needed to gesture toward the nature and consequences of biocultural evolution in our lineage.

A Million Years of Music has drawn sufficient attention to encourage me to elaborate and generalize the biocultural evolutionary patterns that I described while formulating my musical arguments. These patterns of our emergence take center stage here. *Culture and the Course of Human Evolution* addresses a broader set of issues than *A Million Years* did, and it deepens and solidifies their theoretical foundation in the effort to build an encompassing model of the emergence of humanity in its modern form.

CHAPTER 1

INTRODUCTION

:1:

The last major transition

An important initiative in evolutionary theory crystallized in 1995, when John Maynard Smith and Eörs Szathmáry advanced the idea that the history of life on earth had passed through a small number of "major transitions." These were turning points at which decisive new kinds of organization appeared in living things and opened up new possibilities for subsequent evolution. Maynard Smith and Szathmáry saw the transitions not as deviations from the mechanisms of natural selection but as generated by them, enrichments and complications of Darwinian processes. By identifying jumps onto new levels of complexity and postulating the selective mechanisms driving them, "major-transition theory" formalized the idea shared by many scientists that the intricacies of natural selection could sometimes move in ways other than gradual, small-increment change. Another version of this idea in the years before 1995 had been Stephen Jay Gould's "punctuated equilibrium," long periods of stasis in the evolution of species interrupted by quick change and diversification. We can think of Maynard Smith and Szathmáry's transitions as very large, important punctuations.

The major transitions mark pivotal changes in the nature of life (Maynard Smith and Szathmáry 1995, 2000). They include the appearance of information-bearing RNA and DNA molecules, cells bounded within a membrane or wall, cells with nuclei and internal organelles, multicellular organisms, sexual reproduction, animal sociality, and, most recently, modern *Homo sapiens*. How pivotal this last transition was! After billions of years of evolution and many fundamental changes in the organization of life, something unprecedented occurred, a change that would, in a few dozen millennia, enable a single species to populate the whole globe, dominate all its ecosystems, genetically engineer plant species through thousands of years of cultivation, overwhelm or domesticate all other large vertebrate species, alter the planet's climate in fundamental ways, and finally set in motion a mass extinction. What made this last major transition so powerfully different from those that had preceded it, and how did it come about?

Two features reappear regularly in Maynard Smith and Szathmáry's transitions. First, they often involve a change in the *target of selection*, that is, a change in the entities on which natural selection operates. To exemplify this, we might think of the first development of multicellular organisms. In a world of single-celled life, about five-sixths of the history of earthly life, selective forces differentiated individual cells (or kinds of individual cells) from one another according to their fitness; *this* bacterial variety rather than *that* had greater fitness in a selective environment. Within an organism, however, selection and competition among individual cell types—our liver cells versus our skin cells, for example—would prove disastrous. Selective pressures on individual kinds of cells had to be somehow suppressed, operating instead to differentiate whole organisms. The target of selection had shifted from cells to organisms. The second feature that recurs across many of the major transitions is a shift in the nature of *information transmission*. The advent of sexual reproduction, for instance, involved processes of meiosis and fertilization that allowed for the shuffling, or "recombination," of genetic material, typically from two parent organisms, into a new amalgam. Compared with the various nonsexual reproductive modes of organisms, this represented a new organization in the passing on of genetic information from one generation to the next.

Biocultural evolution

These two features were also important in the appearance of *Homo sapiens*, the most recent major transition; here too the transmission of information and targets of selection changed. In this case, however, they changed under the aegis of a new force that formed them in ways different from the other transitions. To this force we can assign a name: *culture*. And the kind of evolution in which culture played a role, marginally evidenced in some nonhuman life but hugely consequential in our lineage, is called *biocultural* evolution.

The book that follows is devoted to explaining the uniqueness of human biocultural evolution, and it might be regarded as a long footnote to Maynard Smith and Szathmáry's identification of our evolution as a major transition. Why is a book-length addendum necessary? The unvarnished answer is that scientists thinking about human evolution—or *hominin* evolution, to broaden the focus to include our closest extinct relatives—have not taken full account of the dynamics of culture. Putting culture in its proper place alters substantively our models of hominin evolution and so alters our picture of our evolutionary history, especially in its later stages, when culture loomed large.

I am not suggesting that scientists have not taken *any* account of culture in our evolution. It is no news among evolutionists that it is futile to try to separate off, in hominin evolution, our biological and cultural aspects, or that nature and nurture have developed in tandem with one another for millions of years in our lineage. (And also, arguably, in a number of other lineages of cultural mammals and birds, if with less dramatic results. Wondering why the runaway developments of the hominin lineage did not happen in these other lineages is a salutary exercise that points up the fact that there were selective paths other than snowballing cultural development along which these other species could achieve their own superb adaptive fitness.) But if evolutionists know that culture and biology cannot be separated, they have been slow to appreciate the full power of culture in the biocultural mix.

Human culture is something humanists and humanistic social scientists (for example, cultural anthropologists) are trained to interpret, discerning and differentiating its specific modes and, from these, arriving at its more general forms. But few scientists are so trained, and it should not surprise us that the premises about cultural dynamics they bring to the study of biocultural evolution are often less nuanced than they need to be. Add to this the proclivity of most scientists for quantitative-leaning explanations, so different from the humanistic, interpretive modes arguably essential for understanding culture and its patterns, and the case mounts that humanistic method needs to take its place at the heart of models of human evolution.

: 2 :

Resolving the sapient paradox

This book does not take in all of hominin evolution but focuses on its latest stages, for it is here that the interplay of culture and biology grew especially intense in our phylogeny and, through this intensity (as I will argue), took on qualitatively new forms. I aim to explain how this burgeoning of human cultural capacities could occur over a span of time that is (evolutionarily speaking) very brief. The brevity is in truth dramatic: I am not alone in thinking that the most characteristic and complex features of human modernity were late attainments, not reaching back in their fully modern forms even as far as the advent of *Homo sapiens*. For example, there is much reason to think that the earliest sapient humans did not speak a fully modern language, even though their social communication was already more complex than that of any nonhuman species around us today. They did not create gods or spirits or form religious beliefs, they were not musical creatures in any modern sense, and in other basic aspects of their sociality they did not much resemble us. But then, in a blink of the earth's eye, they were transformed.

What kind of model will it take to represent this transformation? The model will need to offer a new resolution of a problem that has come to be known to paleoanthropologists as the "sapient paradox" (Renfrew 2008; Renfrew, Firth, and Malafouris 2008). The paradox lies in the similarity to us and difference from us of early *sapiens*. By about 200,000 years ago, after a long earlier development, they were physically so much like us, and so different from other early human groups such as Neandertals, that they are considered "anatomically modern humans"; yet they differed fundamentally from us in behavior and, it seems, in behavioral capacity. How could this be so? And how could all the ingredients of modern human behavior click into place without physical transformations noticeable in the fossil record? A satisfactory model must represent this change as quick and even abrupt but require no fundamental shifts in the biological nature of the creatures involved (if many smaller shifts) and no implausible, magic-bullet genes for language, music, religion, and so forth.

Foundations of hominin culture

Notwithstanding the countless evolved differences that entered into the hominin lineage across the last several million years, it is pretty clear that all hominin species have had several things in common, even if they have manifested them in very different form and measure. First of all, ancient hominins were all cultural animals, in a sense of that word that I can already begin to elaborate: they all witnessed behaviors in older individuals in their groups, repeated and learned these behaviors, and passed them on to younger individuals. This *social transmission of learned behavior or knowledge across generations* provides a straightforward baseline definition of culture, one that evolutionists such as Peter Richerson and Robert Boyd (2005) have used very productively.

This definition of culture is broader and also blunter than most of those advanced by anthropologists across the last century, but my project requires such breadth and bluntness. Working to define culture through induction from modern-day or recent historical cultures, as anthropologists have done, cannot help much to shed light on the earliest appearance of culture. The problem can be exemplified by citing almost any anthropological definition. Here is a famous one, offered in 1952 by Alfred Kroeber and Clyde Kluckhohn at the end of a book reviewing such definitions:

> Culture consists of patterns, explicit and implicit, of and for behaviour acquired and transmitted by symbols, constituting the distinctive achievements of human groups, including their embodiment

in artifacts; the essential core of culture consists of traditional (i.e. historically derived and selected) ideas and especially their attached values; culture systems may, on the one hand, be considered as products of action, on the other, as conditional elements of further action. (Kroeber and Kluckhohn 1952, 181)

And here is another, simpler but equally famous, offered twenty years later by Clifford Geertz, perhaps the most influential anthropologist of the second half of the century:

> the culture concept . . . denotes an historically transmitted pattern of meanings embodied in symbols, a system of inherited conceptions expressed in symbolic forms by means of which men communicate, perpetuate, and develop their knowledge about and attitudes toward life. (Geertz 1973, 89)

As sophisticated as they are, these definitions presume what any deep-historical account of the emergence of human culture must instead set out to understand. They can gain no purchase on hominin cultures *before* there were symbols or symbolic cognition—cultures that did not frame "attitudes" toward life or mark distinctions of one social group from another, cultures made by animals without language. I will engage with such anthropological definitions of culture (and also of another important anthropological concept, *ritual*) only in chapters 6 and 7, where I offer an account of human behavior taking on its modern form. These definitions cannot serve as my starting point, however, for that would be to take the explanandum as explained. We must begin with more foundational stipulations.

Richerson and Boyd's definition of culture suggests two of these—additional shared features of hominins that reach far back in their phylogeny. In order for the transmission of learned behavior to happen, they all must have been animals capable of looking to their species-mates—their *conspecifics*, as ethologists call them—and imitating them, thereby adopting the behaviors of those others and in general learning from their social surroundings. In an important book of 1991, evolutionary psychologist Merlin Donald took this imitation, which he termed *mimesis*, as the fundamental trait of a long period of hominin evolution, a period of "mimetic culture" extending from almost two million years ago down to the advent of *Homo sapiens*. Donald's mimesis required special capacities on the part of the animals that performed it: capacities to shape voluntarily mental "representations" of perceived actions, capacities novel in the hominin line that differentiated hominin mimesis from the less voluntary and less self-conscious mimicry of other animals. These voluntary action plans made possible cultural transmission and the imitative pedagogy of cultural behaviors long before language existed.

Such mimesis was founded on another general capacity: the ability to understand in some fashion the intent behind a conspecific's actions. This ability could lead to imitation through a projection of one's own intent onto another individual: "*I* want to crack a nut; *she* wanted to crack a nut; I should take a stone and do it *that* way." This kind of mechanism seems close to Donald's voluntary formation of mental representations. It involves, it is clear, an impressive cognitive feat, one that is not widespread in the animal world today: the sensing of a mind similar to one's own at work in another individual. This capability is known by various terms among scientists today, but most commonly it is called *theory of mind*. Theory of mind comes in several degrees of complexity, and it is probably found among some mammals and birds today in something like the degree to which it was manifest at some very early stage of hominin evolution. In the hominin lineage, however, it eventually outstripped the forms it took in any other. This development was of fundamental importance for what followed in the course of human evolution, and I will have something more to say about it at the end of chapter 4.

For the most part, however, I will not linger at this early stage of hominin evolution. Instead, I focus particularly on the final emergence of our modernity—roughly, across the last 250,000 years—and I aim to illuminate the special dynamics that entered into these final stages. There are two general reasons for this focus, which derive from the successes as well as inadequacies of the earlier literature. On the one hand, the models developed elsewhere for the growth of early hominin capacities—especially theory of mind, mimetic capacities, and social learning, the basic precursors of cultural elaboration—are compelling, and I accept and exploit them. On the other hand, the models do not satisfactorily explain the recent, explosive emergence of cultural modernity among one or a few species of hominins. This explosion marks a final, dramatic divergence of hominin evolution from the evolution of all other lineages, even other lineages of animals that can be thought to have cultures. And this is the heart of the problem of our emergence, the essence of the sapient paradox, and the feature that makes late hominin evolution fatefully and fundamentally *different*.

: 3 :

Coevolution and niche construction Understanding the burgeoning of our modernity requires revising and enriching the available models of the place of culture in our evolution. In doing this I will not depart, any more than Maynard Smith and Szathmáry did, from the basic mechanisms of the evolution of life: inheritance of traits, variation in them, and selection among the variants. Instead, I attempt to understand new dynamics that emerged from the fundamental Darwinian mechanism of inheritance-variation-selection under the special conditions of late hominin cultural development. In this too my account is related to major-transition theory, though it discerns, in this last transition, cultural forces at work that differ from those described by earlier theorists.

To construct my model of hominin evolution I will move along two paths. The first, pursued in chapters 2 and 3, describes and elaborates a dynamic of biocultural evolution that biologists have explored over the last forty years or so. Though much of the material in these chapters will be familiar to evolutionary scientists, I introduce into it a new mechanism and new emphases that point toward the cultural model described in subsequent chapters. The key concepts here are *coevolution* and *niche construction*, terms that are associated today with novel extensions of Darwinian natural selection theory—the so-called "extended evolutionary synthesis" (Laland et al. 2015; Pigliucci and Müller 2010).

The term *coevolution* has been used to refer to several processes, but it most often names the fact that organisms evolve in relation to other organisms, so that the fitness of one depends on the fitnesses of others around it. A famous instance is an evolutionary "arms race" between predator and prey species: as the one evolves more effective defenses, selection on the other leads to more effective weapons—and vice versa. In this basic form the concept is not truly an extension of Darwin's thinking; it is right there in *On the Origin of Species* and *The Descent of Man*, as we will see. But it has served as a starting point for Darwinian extensions in several directions.

Niche construction is connected to coevolution, but it is more inclusive in the relations it recognizes. It refers to the consequences of the fact that organisms and populations of organisms alter their environments even as their environments exert selective pressures on them that determine their reproductive fitness; as the environments are altered, so are the selective gradients. Niche construction in principle includes the relations of an organism to all the aspects of its environment, living and nonliving. Evolutionary biologists who first described biocultural evolution thought of it as a "dual-inheritance theory," with inherited information passed along both genetic and cultural lines; today's niche constructionists, in contrast, envisage a triple inheritance, adding the constructed environment as a third informational channel.

Coevolution and niche construction both name cyclic systems or networks in which the elements are bound in feedback relations: a change in one species alters the selective circumstances of another, which in turn alters those of the first; or a rebuilt element reshapes the environmental impact on the species that builds it. My elaboration of this feedback model includes the effects of *feedforward*, a mechanism exerting controls on feedback systems and distinct from feedback, whether positive or negative (see chapter 2). The inclusion of feedforward dynamics of culture and its patterns will be central to my account of the emergence of modern humans.

A historical science I have said that much of the material of chapters 2 and 3 will be familiar to evolutionary scientists, and we might be tempted to hold hypotheses about feedback systems, coevolution, and niche construction to be self-evident. The recognition of such mutual impact between species, or between species and their environments, seems an obvious enrichment of simpler views of selection in which environments function as stolid and unchanging bulwarks, with changeable organisms knocking up against them and surviving or perishing. But to call this enrichment uncontroversial among scientists would be naive. Niche construction in particular is closer to cutting-edge debate than to settled doctrine; changing environments of the sort it highlights have posed a problem for formal models of natural selection ever since they were pioneered in the 1920s, and its multidimensional networks of feedback interactions in particular make quantification intractable. The reasons for resistance to these concepts, a recent overview of evolutionary thought has argued, are related to the persistent desires of some researchers to quantify evolutionary processes in the manner of harder sciences—a "physics envy" that a humanist might attribute in turn to inadequately taking stock of the fact that the study of evolution is a *historical* study (Pigliucci and Kaplan 2006, 61).

It might well surprise that humanist observer that the consequences of the historical nature of evolutionary study continue in some quarters to be underestimated, even though this nature was signaled in powerful, prescient ways already by Darwin and has occasionally been accorded detailed attention and analysis (e.g., Nitecki and Nitecki 1992)—and even though the historical nature of biology is regularly acknowledged by biologists of all stripes, as in this opening sentence of a recent college text on cell biology: "Biology is based on the fundamental laws of nature embodied in chemistry and physics, but the origin and evolution of life on earth were historical events" (Pollard and Earnshaw 2008). In what follows I take the histories of niche-constructive feedback systems as a starting point, if one that is not fully quantifiable. For, if any species can be thought to have altered their environments enough to have changed significantly the course of their own natural selection, these would have to be human and related prehuman species. Dramatic examples of hominins altering their lived environments include their control of fire, on the order of a million years old, or, more recently, their hunting to extinction of Upper Pleistocene megafauna from Australia to the Americas or, more recently still, the turn to agriculture early in the Holocene in several regions of the world.

Cultural traditions

Examples such as these also suggest *why* hominins could reshape their environments so radically, and this has to do with their advanced technologies, including exploitation of resources around them as tools. To enter on this technological terrain is to encounter again culture, and it enables us to take another step forward in elaborating this central concept. The toolmaking skills attested in the archaeological record over tens of thousands of generations of many species of hominins were learned and passed along through those generations. There were no toolmaking genes to guide them, but instead, in the context of more general, genetically determined capacities, mimetic and pedagogical traditions in which skills and knowledge were reiterated and also altered, at first gradually and sporadically, later in accelerating fashion.

Such *traditions of learned bodies of knowledge and practice* form a basis for any advanced animal culture. This idea is in essence only a slight extension of the definition of culture offered by theorists Richerson and Boyd, while the traditions themselves have been lengthily pondered by many archaeologists and evolutionists—among them philosopher of biology Kim Sterelny (2012), to whom I owe the image of them as "archives" of cultural materials. That these traditions are a prerequisite for *advanced* culture by no means restricts culture in general to hominins. There are in the world today (and no doubt have been historically) examples of animals outside hominin lineages that have created cultures without them—animals, that is, with transmitted learning but without deep intergenerational archives. Still, there can be little doubt that hominins of at least two species (and probably more) have extended culture by deepening its archives in ways not seen in any other animals.

I have moved quickly here from a foundation of feedback cycles and niche construction, historical in essence, to a low superstructure of culture built on it in the hominin instance. If the effects of this culture on the cycles and niche construction are obvious in myriad ways, they are difficult to measure and almost impossible to quantify usefully, especially across evolutionary time spans. What would it mean to assign a number to the evolutionary impact of a million years of Acheulean hand-ax manufacture or to 10,000 or 100,000 years of Neandertal technical innovations? In both cases the technological patterns are well studied, and it is safe to assume that they generally increased the fitness of the hominins who made them as they faced challenging and changeable environments. The technologies, in other words, altered the ways these people altered their niches. Even in the presence of the huge bodies of data archaeologists have amassed, however, attempts to quantify the alterations or the advantages gained must falter. And what, to turn to *Homo sapiens*, would it mean to quantify the rippling effects of ritual structures in society over tens of millennia? We cannot doubt that there must have been some impact on human niche construction, but measuring it exactly or modeling it formally is a project of limited promise. A historical science demands not only the exploiting of formal models and mathematical tools but also a recognition of their limitations and of the potential of other, qualitative methods.

: 4 :

Information and semiosis The second path I have followed in developing my model of biocultural evolution involves widening the theoretical lens through which we view culture to include information and semiosis. This will enable us to understand better the conditions and constraints under which hominin cultures—or any animal cultures, in fact—emerged. And the widened view helps also to explain the kinds of stages and transitions that advancing hominin culture passed through. For these conditions and constraints differ in kind from the hominin capacities I described before as foundational for culture: imitative capacities, theory of mind, and social learning. They are not conditions of a single species or lineage but instead natural, even ontological prerequisites for culture. A way to picture them is to think of an *enclosure* for culture, the space within which culture can arise. In chapters 4 and 5 I describe four nested realms that define and demarcate this enclosure, one inside the other.

The largest of these is the vast realm of information. I have already employed this term more than once in its vernacular sense, but to understand the extent of the informational realm we need to turn away from this everyday usage to something closer to the conception of Claude Shannon and other early information theorists. More radically even than this, we must grasp Shannon's information in its least anthropocentric form: not as a message sent and received, and in fact not as a conveyance of meaning at all, but rather as a sheer formal correspondence of distinct entities in the world that brings about certain effects. The information "in" or "conveyed by" DNA and RNA in producing protein molecules is information of this sort, a matter of causal correspondences of separate entities (nucleotide sequences and amino acids, in simplest form). This molecular system of information can only loosely be said to have meaning—loosely, and by certain late-evolving organisms (us) whose large brains have given them an implacable tendency to find meaning everywhere they look and understand. But we can see from the example how huge is the extent of the informational realm. In this correlational sense, information extends throughout the cosmos and requires the cognizance of no life-form—or indeed any life at all.

Within this immense realm, a small precinct is reserved for correspondences between things or actions that are *perceived and attended to* by living organisms. Such perception is a capability of only a tiny number of living things (relative to the number of all life-forms), all of them endowed with highly developed central nervous systems. In their perceptions of correspondence they bring meaning into the world by making one thing a token, or *sign*, of another; their perceptions create a *semiotic* realm. Semiosis, in other words, is the wrinkle of aboutness certain organisms make by perceiving connections among things; it is the origin of content and of representation, in the sense of one thing manifesting or re-presenting another. Though semiosis exists in creatures without culture, it is foundational for the cultures of all animals that possess them. How far signs extend in earthly life is a matter of lively debate, but undoubtedly they reach far beyond the human. And, as we will see, they fall into analyzable types, more and less complex, which seem to emerge as a product of the increasing complexity of the nervous systems that make them. My analysis involves a group of terms theorized by philosopher Charles Sanders Peirce: sign, icon, index, symbol, and—most important of all—*interpretant.* Carefully defined and brought to bear on our evolutionary past, Peirce's system puts our understanding of both animal and hominin cultures on a new footing, different even from the one proposed by biosemioticians who have eagerly pursued a semiotic course.

Culture and system

If we zoom in further now on the enclosure, enlarging the semiotic realm until it fills the field, we see within it another, much smaller realm: the realm of culture itself. All animal cultures arise from the perception and manipulation of signs. The transmission of learned behaviors across generations is nothing other than a transmission of meanings in the niche of a given cultural species, and these meanings are generated by linking things or actions to one another as sign and object. Some of the capacities that enable a species not only to make signs but also to elaborate such semiosis into culture have already been noted for the case of hominins, and they probably extend in some measure to all cultural animals. Social learning and imitation seem crucial here, as does at least a rudimentary theory of mind; beneath them, well-developed memory systems must be involved. Semiosis does not, however, inevitably burgeon into culture, and countless noncultural animals perceive one thing as a sign of another and act accordingly—that is, they act semiotically. But in species possessing those other capacities, bare-bones semiosis will be fleshed out into cultural semiosis, and learned signs will come to be transmitted down through the generations.

Zooming in once more, we come to the smallest realm of all, the tiny precinct within the cultural realm of *systematic* culture. Here the most complex kinds of signs come into view. They are not merely independently connected each to an object but instead rely for their meaning on their places in arrays or constellations of signs. The words of human languages are signs of this sort. These signs form systems, and they afford, in the species capable of such perception and cognition, the possibility of a systematized cultural transmission in which not only signs but the arrays themselves are passed on. It is hard to say how extensive such systematic culture is or has been among nonhominin species. In a simple form it seems to be witnessed in certain bird cultures today, among some cetaceans and primates, and perhaps in some other instances. Once we have come to understand the nature of the semiotic systems from which it arises, however, it will be clear that in complex forms systematic culture has been found only in the hominin clade among earthly organisms.

Emergent cultural systems and epicycles

Understanding the nature of systematic culture fills in a large lacuna that opens in most accounts of hominin evolution, even those building on the niche-constructive foundation and the cultural first floor I erected before. This is because cultural systems enter into niche-constructive feedback in different ways and with different effects than nonsystematic culture can do. We must understand ancient human cultural systems in several regards: what elements they comprised, what different kinds of them arose, how they came to operate as systems, and what unexpected effects their operation brought about.

These extraordinary effects form important elements in the picture of late hominin evolution I will offer. In the first place, the networked, semiotic relations of the component parts of cultural systems—ideas, behaviors, and material things constellated in mutual interrelations—granted them new, emergent functions made possible by their systematicity. Signs acting in systems, simply put, can have different effects from those acting independently. Second, because of similar general constraints acting on separate groups (even different species) of hominins in their respective habitats, the systems and their emergent functions could come to be highly likely outcomes in the lives of unconnected populations. That is, similar systems were liable to arise independently across a set of independent selective terrains bearing similarities to one another. Third, the networked, organized structures of cultural systems gave them a degree of self-maintenance, robustness, and stability in the face of ongoing selection acting on the species that created them. These self-conserving tendencies resulted in an unexpected positioning of the systems in the feedback cycles of niche construction, granting them *a measure of autonomy from the feedback processes that fostered them in the first place*. The cultural systems, even though spawned from the feedback cycles, come to stand somewhat apart from them, maintaining their integrity while feeding *forward* into them and changing the dynamics of niche construction. Because of this feedforward quasi autonomy, this stepping outside the feedback cycle, I have termed such cultural systems *epicycles*. The cultural epicycle is a concept I introduced in an earlier book on human evolution (Tomlinson 2015). I elaborate its consequence here.

: 5 :

History and historicity

Cultural systems, in broad perspective, are a manifestation, late in the evolution of life and in one small corner of the animal kingdom, of the emergent complexity scientists have discerned in many other systems, living and nonliving—but especially living. Like metabolic systems, systems that transmit genetic information, and ecosystems, they are assemblages that maintain themselves far from equilibrium through the reciprocal interactions of their elements. They raise basic questions about their self-maintenance and the nonlinear or feedback causal paths they create. And, like those other, biological systems, they pose the additional challenge of understanding the evolved histories that gave rise to them. Their interpretation brings us once again to confront the historical nature not only of the human sciences but of life science all told.

This historical condition, or *historicity*, casts us into the methodological currents and countercurrents that run through the straits separating science and the humanities. Appreciating the nature and evolutionary impact of cultural systems and epicycles requires not merely customary scientific tools—though it cannot do without them—but also the interpretive tools developed in humanistic and historical investigation across the last century. Because such investigation is not the concern of most scientists, the systematics of culture has mostly escaped from their accounts of human evolution. The gap in thinking about hominin evolution, in other words, has opened up in the distance between science and the humanities, and it now calls for some measure of humanistic theory and method to narrow it. Evolutionists, for example, write on information and sign making in our cultural evolution without addressing the unique systematizing of signs that characterizes *Homo sapiens* and perhaps a few other late hominins.

Many archaeologists, meanwhile, see the concept of symbolism as fundamental in determining the coalescence of human modernity, but they seldom pause to define it or discern what kinds of semiotic and cultural systems support it, what general roles these played in human evolution, and what still more general cognitive modes they rely on. The major-transition theory with which I began offers a third case in point. Maynard Smith and Szathmáry were unequipped to explain *why* human evolution could amount to a major transition, though they worked to make a limited view of language and its acquisition fill the bill. Rightly deployed, the systematics of culture might provide the ingredient missing from their account.

The issues raised by evolved and emergent systems finally pose a large question about historicity: What methods will be most productive in trying to capture its operation and effects? The historicized cycles and networks of variables characteristic of cultural coevolution and niche construction are not susceptible to rich analysis with methods that necessitate a reductive thinning-out of evidence and posit one-way chains of causation. These are methods founded in positivism, and the theory and practice of history over the last century has taught us that they cannot provide the answers historical analysis seeks. For evolutionary histories, this applies to local events, such as the emergence of a single species, the appearance of a complex organ or behavior, or the rise of complex cultural systems; and it applies a fortiori to the largest evolutionary changes, such as any one of the major transitions with which I began, including the emergence of human modernity.

I noted above that evolutionary scientists from Darwin on down have not been insensitive to the historicity of their endeavor. To address it, many have turned toward *modeling*—a word that I have used already in this chapter but that calls for closer scrutiny. Through various means of mathematical, graphic, and computational representation, historical processes and change can be visualized, if not tested and falsified; in this way models offer a heuristic that is especially effective in a historical science. In doing so, however, they come to occupy a middle ground between the theories they instantiate and the reality they (in some fashion) duplicate. In relation to theories, as economist Axel Leijonhufvud has pointed out, models are radically incomplete—"formal but partial representations," always leaving out "lots of stuff" (1997, 193). In relation to the historical situations modeled they are also incomplete, hard-pressed to capture multidimensional causality, highlighting one or a few aspects of what they model while neglecting others. Doubly reductive, evolutionary models confront us with the question of how far their formalization can proceed—how rigorously, that is, their components can be defined and the relations of these parts specified—without vitiating their explanatory power. I will take up several sorts of quantitative and graphic models in the coming chapters, and the reader can judge their scope and usefulness.

Descriptive understanding

However useful the models may be, the bottom line is that evolutionary histories, like all histories, will not reveal their secrets on the experimenter's lab table, the mathematician's scratch pad, or the computational modeler's screen. Whatever large contribution those venues offer—a contribution I have no wish to minimize—evolutionary histories require the support of other methods. They must be portrayed, explained, induced from limited examples. And most of all they must be *described*—a very humanistic word, allied with historicity. Description works to fill in the space between models and the realities they purport to mirror, nuancing the one and beckoning toward the other. (Leijonhufvud [1997] recognized something akin to description also in the gap between model and theory, describing a rhetorical, prosaic filling-in that must accompany formal models in order for them to represent theories effectively.) Historically oriented humanists might find superfluous my singling out of description as a method, for to them it is second nature that past circumstances must be gotten at by evidence-supported, nonreductive, uncontrolled, sprawling, and incomplete prose accounts. They are schooled in modes of elaborated description, contextualized and *thick*, as cultural anthropologist Geertz (1973) used to say, developing notions he had found in the work of philosopher Gilbert Ryle. Their approach gains its power and validity not from proof or demonstration, and not from mathematical rigor, but from the conjuncture of two difficult-to-quantify features: its richness of content and coherence of connections.

Although scientists have now and then thought of verbal description as a model of a certain sort (Bertalanffy 1969), it remains a method foreign to many of them, and humanists who converse often with scientists—their number is growing—will have encountered reasons why it needs to be lingered over here and even defended. I remember talking with a distinguished senior neuroscientist who couldn't see the point of description of any sort. Its tautological quality, replacing a real situation with a diffuse facsimile, could not *explain*, since this end could be attained only by tracing causal pathways through empirical investigation. This was for him what all *analysis* is about, and analysis, he maintained, is the path to true and certain *definition* of the objects of our study.

Such reasoning, however, calls into question the explanatory force not only of description but of *all* models, since they are by design and nature always tautological. Mathematical and computational models, thick description, and narration all pose the question of how repeating something in a different form can help to explain it. At the same time, and somewhat paradoxically, they all predicate their explanatory power on the fact that there *can* be no such thing as sheer repetition in a different form. Indeed, in humanistic theory it is now accepted that understanding can arise in the slippage between a description and its object, like the displacement between one view of a thing and another that brings about parallax. Whole disciplines in the humanities—perhaps the humanities all told—arise from this practice of what we might call re-scription. But then, something similar can be seen to be foundational for human language, intersubjectivity of all sorts, and even the structures of the psyche and consciousness. All of these, including humanistic description or re-scription, can be thought of as experiences of slippage.

And we can think of historicity also as a most general experience of slippage, one that is, I will argue, characteristic of *Homo sapiens*. So the project of tracing our emergence doubles over on itself: we must describe a history of our species and other, related ones across evolutionary time, starting from a period when the beings whose history we trace could have had no historicity, and we must discern how certain modes of slippage came to be basic to human experience—and then to shape, necessarily, the explanatory methods we deploy. It is a relief to say that actually describing evolutionary histories, which will be the main business of this book, does not require a full-dress working-out of these methodological puzzles, and it is useful in this regard to recall again the example of Darwin. In all his work, but especially in *On the Origin of Species*, he made it clear that an account of the origins of biological complexity requires descriptive creativity as a part of the explanatory apparatus. This was as true for the various mechanisms of speciation he proposed as it was for the emergence of complex organs such as the vertebrate eye or complex behaviors such as hive making among bees, for both of which he provided plausible narratives. And it remains true for us today in trying to understand the emergence of human modernity. For Darwin as for us, it is the flux or process or *experience* of history, historicity itself, that calls for the descriptive mode.

: 6 :

Archaeology It is precisely historicity—or the effort to capture its symptoms and effects—that makes the anthropological definitions of culture I quoted before inapt as my own starting point. I proceed instead, after building an edifice from the recent developments in extended evolutionary theory and in humanistic theory that I have already summarized, by exploiting archaeological evidence to summon up distant ancestors of ours that were cultural animals but that showed nothing like modern cultural elaborations. That is, my approach does not begin from present-day circumstances and try to use them to model the deep past; rather, it musters evidence to describe the conditions of early humans. My models aim to represent the conditions and, additionally, to gain some modest power to point toward the course of their development—to point, that is, from more ancient conditions to less ancient ones.

This method culminates in chapters 6 and 7, where archaeological evidence concerning the more-than-200,000-year career of *Homo sapiens* comes to the fore. Chapter 6 describes in general terms my model for the biocultural evolution of *Homo sapiens* across the period from about 200,000 to 50,000 years ago. It takes shape from several kinds of evidence: archaeological, genetic, paleoclimatological, and paleodemographic. The model I build from these puzzle pieces indicates that the systematic cultures described in chapter 5 came to play a novel role in our selective evolution across this period. It shows how separate populations of humans could converge independently on certain shared capacities, cultural and (ultimately) biological, and it indicates why the feedback dynamic that enhanced our capacities for cultural niche construction finally reduced selective pressures on those capacities, bringing general change in them to a virtual standstill.

In chapter 7, finally, a closer look at the model reveals in finer grain the shift from near-modern to fully modern humans that occurred within the species-span of *Homo sapiens*. Here we see the forces that transformed humans on the verge of modernity into humans like us in all fundamental ways. Understanding this transformation depends on a full appreciation of the consequences of the heightened interplay of cultural systems in human niche construction and in the reshaping of our selective terrain. In this chapter I model the workings of this interplay as it brought about the emergence of modern technology, ritual, music, language, and metaphysics.

Throughout *Culture and the Course of Human Evolution* the watchwords will be clear. *Niche construction* builds *feedback* loops between the environment and the genome, altering selective terrains for many, probably all, organisms. *Semiosis* arises with the perception of an aboutness in the world that many animals experience. From this springs, in the behavior of fewer animals, *culture*, and where cultures arise they become active forces in niche construction and its dynamics. Still fewer animals construct and transmit culture in the form of organized, hierarchized *systems*; to do so in any advanced form probably requires cognitive capacities limited to the late hominin line, capacities not only for the organizing of cultural forms but also for the sheer *accumulation* of deep archives of cultural knowledge and practice. The resulting systematization enables culture not only to influence niche-constructive feedback cycles from the inside, so to speak, but to assume in varying degrees a controlling, *feedforward* influence from outside the cycles. These watchwords are, in their ways, straightforward; from their interactions, however, emerged dynamics complex enough to bring about the last major transition in the history of life.

CHAPTER 2

THE CONSEQUENCE OF FEEDBACK I: COEVOLUTION

:1:

Feedback and evolution

Feedback is hard to think. The idea that A causes B which causes A looks simple, but its circularity breeds confusion, flying in the face of deep-seated conditions governing how we form knowledge about the world. The challenge of thinking even a simple feedback system confronts us in everyday life. When we turn on our furnace, we understand ourselves to be causing it to start up, but the thermostat mechanism by which it is turned *off* without our assistance once the house is heated is a bit less easy for us to take in. The compounding of interest on credit-card debt is trickier still to grasp—a fact banks exploit to submerge millions of customers in long-lasting debt. Einstein is supposed to have called compound interest "the most powerful force in the universe"; with less hyperbole he added (the story continues), "He who understands it, earns it; he who doesn't, pays it."

The confusion bred by feedback systems arises also in less quotidian pursuits and helps to explain the historical fact that feedback *theories* came late, long after the design of feedback machines themselves. These machines have a history reaching back to the ancient world, where water clocks were equipped with float valves that maintained a constant water level by closing off input as the level rose and opening it as it fell. A more recent feedback mechanism, the centrifugal governor that James Watt designed in 1788, was also used for regulation, in this case of the new and epoch-making steam engine. Here power from the engine spun a system of fly-balls; as they spun faster and rose, they gradually closed a valve regulating steam flow into the engine, damping its performance. Watt's device and similar mechanisms that followed stimulated early moves to generalize their operation, most famously in James Clerk Maxwell's paper "On Governors," delivered to the Royal Society of London in 1868. This represents an important step toward modern control theory (Mayr 1971), and indeed, the conceptualizing of feedback only truly came of age in mid-twentieth-century control, or "cybernetic," theory, dynamic systems theory, and information theory (Wiener 1948; Bertalanffy 1969). These bodies of thought all emerged in the wake of additional technology: the design of electronic circuits and exploration of the practical consequences of "feeding" the output of a circuit "back" into it. Finally, in the late twentieth century, theory and technology were fully joined in digital systems and processing, in a development that has come to be of such pervasive importance that we might justifiably call our time the age of feedback.

The difficulty of conceptualizing feedback is also plain to see in biological theory. Here feedback systems play a fundamental role, and today they are identified or postulated on every scale: from molecular networks governing the expression of genes and cell metabolism, through the life cycles of individual organisms and species, all the way to epochal processes and major transitions in the history of life. But the role of feedback remains hard to pin down, hard to model, and hard to formalize. The situation is clearest in the area of evolutionary theory (Robertson 1991). Feedback operations are basic to ideas that have joined together over the last several decades to form what is sometimes called the "extended evolutionary synthesis," an outgrowth of the "modern synthesis" that had united Darwinian natural selection and Mendelian population genetics around 1930 (Pigliucci and Müller 2010). The novel ideas making up the extended synthesis remain controversial, however, and this resistance has been traced in part to the models of "reciprocal causation" that their feedback operations introduce (Laland et al. 2015). It is revealing that the major study of one of these ideas, niche construction, dates from as late as 2003—and even at that date could justly be subtitled *The Neglected Process in Evolution* (Odling-Smee, Laland, and Feldman 2003).

Darwin and feedback

There is an irony in this situation, because feedback mechanisms were basic to the hypotheses of evolutionary theory from its start and are prominently featured in Darwin's *On the Origin of Species* of 1859. The basic dynamic of natural selection itself shows elements of feedback design. Organisms inherit their traits from previous generations but also vary them, and this variety is either culled by environmental constraints or thrives in them. If it thrives, environmental influence in effect feeds back onto a population so as to increase the frequency of the trait in successive generations, and this, under circumstances Darwin spent much of his book proposing, might eventually lead to new species. An even clearer feedback system in the *Origin* involves the interplay of one species with another. The grand "entangled bank" with which Darwin ended his book was the result of a "struggle for existence" whose mutual balances were determined by natural selection. Darwin envisaged a give-and-take among the different organisms in any ecosystem in which, "if some of these many species become modified and improved, others will have to be improved in a corresponding degree or they will be exterminated" ([1859] 2003, 620). Today this mutually adaptive entanglement of species travels under the name *coevolution*.

If we look to the background of Darwin's vision, we find another feedback system. His inspiration in 1838 for the struggle for existence, and the starting point for his exposition of it later on in the *Origin*, was Thomas Malthus's *An Essay on the Principle of Population* of 1798 (Malthus 2008). Malthusian population dynamics is itself a feedback design, built on the mutual interaction of population size and resource availability. In summary form: A population increases faster than the availability of its resources, resources per capita decline, the underresourced population declines, and resources rise again. A changes B which changes A which changes B. And the connection of natural selection to feedback mechanisms was more than a buried implication in the nineteenth century. In the now-famous paper he sent to Darwin the year before the publication of the *Origin*, Alfred Russel Wallace, the codiscoverer of the theory, hypothesized a mechanism by which the varieties of living things must diverge ever farther from their original types. Wallace's mechanism ensured a balance of organisms' capacities with their needs and the resources of their habitats, and in this, he wrote, it operated exactly like Watt's self-regulating centrifugal governor (Wallace 1858).

Feedback is nothing new to evolutionary theory, then. That it characterizes the basic dynamics of both adaptation and population flux was understood by their first theorists. Today, however, the new ideas of the extended synthesis of evolutionary theory have expanded and enriched the roles of feedback as never before. It is the task of this chapter and the next to review and in some measure elaborate these theories as a foundation on which to build a new view of human evolution.

:2:

The modern synthesis

The modern synthesis of evolutionary theory took shape across the first four decades of the twentieth century and merged the two greatest achievements of nineteenth-century biology: Gregor Mendel's genetics and Darwin's natural selection. It can come as a surprise to learn that these achievements, so fundamental now, languished for decades after their first formulations. Mendel's analysis of pea plants in the years around 1860 showed that the traits of individual plants could be tracked through generations as the interplay of discrete, particulate units of inheritance;

but his findings were largely ignored until 1900. Natural selection, on the other hand, was certainly *not* ignored, but it suffered in a different way: exactly by virtue of its proliferation. Darwin's Malthusian struggle for existence was one of those ideas irresistible in its moment, and it came, across the decades after 1859, to be dispersed into all areas of popular doctrine, providing an ideological umbrella under which ideas scientific and social, reputable and reprehensible, could gather. The proliferation of the idea was accompanied by a dilution of the power of Darwin's specifically biological conception, a dilution perhaps inevitable as long as there was no discernible unit of inheritance and variation.

The scientific rehabilitation of both ideas began when they were joined together in the years after 1900; this was the beginning of the modern synthesis (Huxley [1942] 2010; Depew and Weber 1997). Mendel had linked the traits of individual organisms (by now called their phenotypes) to their particulate units of inheritance (their genes, collectively making up their genotypes). His finding was now extrapolated mathematically to whole populations of organisms, in analyses of the varying *frequencies in populations* of distinct phenotypes and the genotypes they "expressed." Particulate inheritance met statistical analysis, and population genetics was born. The analysis of change in gene frequencies across generations immediately suggested the relevance of the new field to evolutionary theory, and Darwin's natural selection came into a new, sharp focus. The three ingredients of his algorithm had always been clear: *inheritance* of traits, *variation* of traits, and unequal *selection* of the variants. Now it was understood that variations in genotypes (mostly through mutations) offered up new individual phenotypes to environmental constraints and that, as these variants were differentially selected, the frequency of traits in the whole population would change. Here, finally, the motor of evolution through natural selection came into view, the genetic means by which populations of organisms could change over time: varying, thriving, dying off, and—most important—diverging into new species.

This synthesis of natural selection theory and population genetics came into its first maturity in the years between World Wars I and II. From the 1940s on through the next several decades, its powerful implications were explored by the major figures of genetic and evolutionary biology. These are the years of giants in the field: Ronald Fisher, J. B. S. Haldane, and Sewall Wright at the beginning of the period, Theodosius Dobzhansky, Ernst Mayr, George Gaylord Simpson, and John Maynard Smith later. Along the way, in 1953, the synthesis received the most dramatic kind of support with the discovery of the structure of the genetic molecule, DNA, by James Watson, Francis Crick, Rosalind Franklin, and others. If Mendel's work had revealed the motor of natural selection, now its operation was known in new detail, and molecular biology leapt forward with a momentum that has not slackened today.

Selfish genes and memes

This momentum brought a crescendo of acclamation for the centrality of the gene in evolution: the so-called "gene-centered" view. The loudest support came from Richard Dawkins, who in the 1970s offered an account of genes as "selfish" replicators that use the bodies of the organisms they inhabit as "vehicles" or "survival machines" and exploit metabolic energies and resources for their own propagation (Dawkins 1976). Even the suite of an organism's actions in the world—what Dawkins (1982) would call its "extended phenotype," as seen, for example, in a termite mound or beaver dam—comprised first and foremost mechanisms ensuring the survival and replication of the genes controlling the actions.

Pushback against Dawkins's extreme position was quick to form, voiced by Richard Lewontin, Stephen Jay Gould, and others (Lewontin 1983). They did not dispute in any general way the consequence of genes in evolution but aimed to modify Dawkins's position by asserting the consequence of phenotype and environment *in addition* to genes and by broaching the complexities of interaction among all three (Lewontin 2000). It is not too much to say that all the major moves in the extension of the modern synthesis of evolutionary theory have concerned these interactions in their various forms and at various levels of scale, from molecular to ecosystemic. And, since the potential for feedback relations in these interactions is a rich one, the debates over the gene-centered view prepared the ground for advances in this area also.

One other aspect of Dawkins's gene-centrism needs to be taken up here, since we are concerned with the quintessential cultural animal, *Homo sapiens*. Dawkins carried the model of gene action over by analogy to human culture. He proposed that culture has its own replicators, analogous to genes in their particulate nature, their transmission, and the action on them of selection. He called these replicators *memes* (Dawkins 1976)—and memes, like Darwin's struggle for existence a century before, proved to be an idea irresistible to popular culture and popular science alike. The term spread quickly through cultural analysis, giving rise to the ersatz field of "memetics" and becoming the term of art to signal every recurrence of a cultural gesture, every sign of a cultural connection, and finally almost any cultural move that we *recognize*: the Nike swoosh, the Coke *C*, connections between two hit songs, the repeating themes of a political season.

But if the word is by now lodged in all manner of public parlance, the concept remains problematic. It is important to identify exactly where the difficulty lies. It is not found in memes' signaling of the transmissible nature of culture. This is a given of all human culture, which, complicated and hard to analyze as it is, does not need memetics to emphasize it. Neither is the problem the idea that *cultural* selection might take place along lines analogous to *natural* selection. Dawkins borrowed this insight from population geneticist Luigi Luca Cavalli-Sforza, and, as we will see in chapter 3, it has stimulated well-developed positions in the extended evolutionary synthesis. The problem with the meme concept is not even the idea that discrete, exactly duplicated gestures play an important role in culture. In itself, this idea is unassailable, as there is and always has been plenty of more or less exact duplication in human cultures. Those swooshes and *C*s may exemplify modern, precise mechanical and electronic reproduction, but we can trace their kind of gestural repetition all the way back to the design resemblances of Lower Paleolithic hand axes a million years ago.

The problem of memes concerns, instead, an assumption that is built on all three of these premises: the notion that culture evolves according to the Darwinian algorithm of selection *because of* the particulate nature it shares with genes. At a time when even the particulate nature of genes has been blurred from various quarters of developmental and molecular biology, memetics advances an idea of cultural development and history as a flow of discernible, delimitable units. Every attempt to define these, however, runs up against the proliferative, nonparticulate tendencies of human sociality in all its modes (Bloch 2000). The Nike swoosh may be repeatable, but it is fundamentally unlike a gene for a pink petal on a pea plant, since it is no singular gesture at all but always both the product of and an agent in a changing network of cultural meanings. And it is the place of a duplicated cultural gesture in the network that is crucial in making it meaningful. (For Terrence Deacon [1999] this reveals memetics to be merely a schematic simplification of the study of the systems of signs that found all cultures.) Genes, however complex and elusive their identities, transmit their effects though a conservative system in which nucleotide sequences are copied faithfully in the vast majority of their replications, miscopied rarely. Cultural gestures, even those ostensibly reproduced exactly, enter onto a terrain of meaning where they differ in each new instance—even while looking identical to the memeticist, even in those cultural settings that the anthropologist or sociologist might call "conservative." To *isolate* a meme's duplication, then—the chief aim of most memeticists thus far—is exactly to render it meaningless, to cancel out all the meanings that bubble up, fade away, and change with its repetition, and therefore to set aside all the reasons why we might wish to *understand* its duplication. Cognitivist and anthropologist Dan Sperber, long an opponent of memetics and the notion of culture as replication, has summed it up this way: "A process of communication is basically one of transformation" (1996, 83). This is a truth about human culture that is not limited to its modern forms but reaches back dozens of millennia, at least.

That memes ultimately offer a *reductivist* view of human culture, its transmission, and its evolution is recognized today even by those thoughtful students of cultural evolution who wish to retain Dawkins's concept in qualified form (e.g., Henrich, Boyd, and Richerson 2008). Memes point the way not to a cultural science but to a cultural scientism, a method for generating thin descriptions; and in their reductivism they are akin to a problem that we will identify in other, more serious scientific approaches to culture. The alternative developed in this book is the amalgam of science, humanistic theory, and historicism introduced in chapter 1.

: 3 :

Coevolution

In the wake of the gene-centered view came the extended synthesis in evolutionary theory, with its expanded roles for feedback mechanisms. The first aspect of it I will take up is coevolution: reciprocal relations between the evolutionary adaptations of different species. Darwin had already posited this in *On the Origin of Species*—for example, in this meditation on clovers and bees:

> The tubes of the corollas of the common red and incarnate clovers . . . do not on a hasty glance appear to differ in length; yet the hive-bee can easily suck the nectar out of the incarnate clover, but not out of the common red clover, which is visited by humble-bees alone. . . . Thus it might be a great advantage to the hive-bee to have a slightly longer or differently constructed proboscis. On the other hand, I have found by experiment that the fertility of clover greatly depends on bees visiting and moving parts of the corolla. . . . Hence, again, if humble-bees were to become rare in any country, it might be a great advantage to the red clover to have a shorter or more deeply divided tube to its corolla, so that the hive-bee could visit its flowers. Thus I can understand how a flower and a bee might slowly become, either simultaneously or one after the other, modified and adapted in the most perfect manner to each other, by the continued preservation of individuals presenting mutual and slightly favourable deviations of structure. ([1859] 2003, 611)

The feedback pattern in Darwin's example is clear. It is advantageous for clovers to be pollinated, advantageous for bees to suck clover nectar. Natural selection will therefore favor bee variants with enhanced ability to gain access to the nectar, and clover variants making nectar more accessible to the pollinators. Through two separate vectors of selection, two species come to be linked as determining elements in each other's selective environments. If there is a perturbation in the local ecology—here, the disappearance of the humble-bees—the selective pressures on the clovers and the remaining hive-bees will be adjusted. The mutual adaptation of the species, their coevolution, will continue in a reshaped way.

In its classic usage, the term *coevolution* refers to this interlinked, feedback relation of the selective histories of species, groups, or populations. These can be, as here, histories of mutual advantage, but they need not be. The relations of predator species and the species they prey upon form coevolutionary pairs. As enhanced modes of attack (or defense) are selected in one species, variants in the other species are selected that alter it toward more effective modes of defense (or attack). This is the well-known evolutionary "arms race." Darwin saw this possibility too and described it in a hypothetical example of wolves and deer that came just before the clover-and-bee example in the *Origin*.

Symbiotic relations of all kinds also involve coevolutionary feedback. The impact of such symbiotic coevolution has, by all recent accounts, been immense in the history of life. To cite two major, early episodes of it, now generally accepted: both the mitochondria found today in most complex cells and the chloroplasts found in photosynthetic cells are the remnants of the invasion of ancient cells without such organelles by other, single-celled symbionts. The subsequent coevolutionary history—by which, for example, the DNA, RNA, and proteins of the host cells came to be linked metabolically to those of the eventual mitochondria—must in each case have been long and complex.

Optima networks; the Red Queen Of course, groups of organisms in the real world, whether species or other taxonomic levels, are not linked to one another in a one-to-one, exclusive relation. Ecosystems are entanglements (to recall Darwin's word) of countless groups, mutually bound in beneficial or harmful ways, and the connections of their selective histories extend promiscuously in many directions. For any individual population of related organisms we can picture the lines of coevolutionary connection as forming a network of selective relations with other groups, a network that extends and changes across evolutionary time. Bees compete for nectar with other insects and birds, and they fight off microorganismal parasites, predators for which they are prey. Clovers vie with many other plants around them for resources in the soil or space in the sun; they must survive being grazed by passing herbivores and might depend on those animals for dispersal to new terrains. All these relations are not static but change with each perturbation of the ecosystem: a new herbivore, a more aggressive microorganism, a fungus that kills off a rival plant.

In the large scheme of any ecosystemic entanglements, most coevolutionary relations will not be mutually beneficial in the happy manner of clovers and bees. This seems intuitively right, since all the organisms in an ecosystem are competing to enlarge their share of the limited energy bound up in it. Darwin's struggle for existence is on the whole truly a struggle, not a coalition for mutual advancement. In 1973 the evolutionist Leigh Van Valen gauged this struggle against the paleontological evidence of extinction rates for many groups. He was puzzled that, in successions of evolving life-forms, later ones did not persist longer than earlier ones, as we might expect if the successions reflected adaptations making for ever-improved fitness in the relevant "adaptive zones" or ecosystems. The mistake, he saw, was to think of the zones as static, when in reality they are "resource spaces" constantly depleted by the groups struggling to survive in them. In any adaptive zone, then, each group is faced with a similar constant depletion of the resources to be won, and its position cannot be secure. The ecosystem as a whole is "an ensemble of mutually incompatible optima," each one the optimum for a given species or group, and each group struggles to gain and maintain its own optimum. Natural selection in this view does not result in a constant improvement of organisms in relation to their ecosystem but is more like a constant movement not to slip away from an optimum on an ever-mobile terrain. Van Valen (1973) pictured each group running to maintain its place, and, with a bow to Lewis Carroll, he dubbed his model the "Red Queen's Hypothesis."

In this assemblage of competing optima, feedback relations remain paramount. They mutually shape selective pressures in the many species or groups bound in an ecosystem, favoring this phenotype over that in each group and thereby gradually altering its identity. But the alterations only keep groups in the running with the alterations of all the other groups connected to them. Coevolution is nothing but this network of mutual change in interacting populations, and feedback relations extending across many generations have proved to be a fertile way of modeling ecosystems and species histories.

Levels of coevolution

I have been circumspect about specifying the exact level in taxonomy at which the feedback operates. Is it the species level? A higher taxon such as genus or tribe? A lower one such as subspecies or variety? My caution is due to the fact that the relations must be conceptualized as working at various levels; different feedback systems will come in and out of focus according to the more or less fine grain of our investigation. To return to Darwin's example, we can understand as he does the feedback operating between hive-bees and red clover, or we can think about its operation between two types of bees and two types of clover. At this level too feedback systems extend across evolutionary history.

This fluidity of levels signals a broader, finally pervasive operation of feedback systems in biological organization. In recent biological theory, feedback is discerned at all levels in the organization of life and its history. For example, autocatalytic systems of molecules—molecule A catalyzes the production of molecule B which catalyzes C which in turn catalyzes A—are proposed as fundamental to the origin of life (Kauffman 1993; Maynard Smith and Szathmáry 1995). The conservation of basic body designs and features across huge stretches of evolutionary time is traced to self-maintaining networks of genes and proteins, stable feedback systems which in turn stabilize the unfolding development of organisms and resist radical alterations in it (Wagner 2014). At the opposite end of the organizational scale, ecosystems cannot be well understood except as flows of matter and energy through feedback "hypercycles"—seen, at the most basic level, in food-chain cycles in which dead matter provides sustenance for organisms successively higher on the trophic scale, which die and reenter the cycle. These ecological feedback cycles operate on all levels, from the local to the planetary (Wilkinson 2006).

Intraspecies coevolution

Even if we narrow our focus back to the coevolutionary relations of two species only, feedback reveals multilevel complexity. Across generations a predator and its prey are linked in a coevolutionary relation, but what kinds of features determine which *individuals* in the predator population will be most successful in hunting or which in the prey population will be most vulnerable? Several features come quickly to mind, and some of these will be reflections of ontogeny, not a selective, phylogenetic history, as when very young or old individuals in herds of herbivores are targeted by carnivores. However, in many animal species the foremost general phenotypic distinctions are sexual traits, and these differences too can be linked to the feedback arms race. The peacock's tail is an extraordinary display to attract the peahen, but it also requires matter and energy for its development and slows down its bearer, making him more vulnerable to predators. It is selectively advantageous only so long as the resources it requires and the disadvantages it presents do not outweigh the preferential reproductive access it gains. This means that the *inter*species feedback (predator and prey) has inserted itself into *intra*species (sexual) dynamics.

Moreover, the sexual dynamic of the peacocks and peahens in itself could have arisen only through another selective feedback process—this time in the subprocess in natural selection that Darwin ([1871] 1981) described in his second great book: sexual selection (see also Prum 2017). The magnificence of the peacock's tail and the sexual displays and behaviors of many other species seem so natural that we sometimes forget their own selective histories, which involve an arms race at a level different from predator and prey: the level of competition among individuals of the sex doing the displaying (usually male, but not always). To speak only of birds: bright plumage, elaborate songs, and spectacular mating dances are all costly behaviors, and the cost must be balanced by a gain calculated across evolutionary history. The calculation has to do with swaying the nondisplaying sex, of course. Gaudy tails do not attract females by some universal rule and as if by magic; there needed to be a reciprocal selection between the male and female birds that selected not only grander tails but also females that noticed the grandeur.

The history of coevolution, in this way, turns out to be one of loops networked with other loops at each level and intersecting with the loops of other levels. In other words, feedback all the way down and all the way up.

: 4 :

Positive and negative feedback

This survey of coevolution puts us in position to say something more about the general nature of feedback. Feedback systems come in two varieties, negative and positive. In negative feedback systems, the output that is fed back into the circuit opposes what generates it; this is the variety that is represented by Watt's governor (increased engine activity throttles down steam input) or that turns off your furnace once the ambient temperature reaches a certain level (here there is a threshold at which the rising room temperature, gauged in your thermostat's thermometer, breaks the electrical circuit and turns off the furnace). Negative feedback can bring about overall stability in a system—homeostasis, in other words—if the contrary effects are brought into balance. Homeostasis will also often involve oscillations, as happens with your furnace: once the room temperature falls back below the threshold, the thermostat closes the circuit and turns on the furnace.

The balances of the natural world, from bacterial colonies to the largest and most complex ecosystems, pervasively involve negative feedback. Not only do bacteria grown on agar come up against resource limitations as the population increases, but their colonies also conform to spatial limitations determined by the diffusion of resources through them and even produce their own metabolic inhibitors of growth (Cooper, Dean, and Hinshelwood 1968; Hochberg and Folkman 1972). Among animals, a predator population is limited by the evolved defenses that make its prey more elusive, and the prey population is limited by the effectiveness of predator hunting (as well as by other factors, such as the availability of its own foodstuff). Malthus's population dynamics, so important for Darwin, involve a negative-feedback circuit whereby population growth is slowed by resource depletion. In taking over this dynamic as the basis for natural selection, Darwin assumed along with Malthus that any population would expand exponentially in a situation of bountiful resources. Since such a situation could not last for long as a population grows, he reasoned, natural selection would operate to weed out the most vulnerable variants in a population—functioning, in other words, as negative feedback.

If negative feedback often helps to bring about the balances of natural systems, positive feedback pushes against its constraint. The very tendency of populations to increase geometrically is an effect of positive feedback, as the population in one generation reproduces more than its number, giving rise to a population whose enlarged numbers become reproducers themselves. The explosive growth that results is not always immediately dampened in local ecosystems, as anyone who has had the flu or a strep throat can attest. Moreover, a negative feedback without any amplification in the system would soon reduce its output to nothing, so in biological systems positive feedback is regularly linked with negative feedback in oscillating cycles. Thus, Darwinian selection, with its negative feedback, is driven by the tendency to population growth. In the Malthusian model, if the human population declines because of the negative impact of reduced resources per capita—say, rabbits and birds to hunt—the rabbits and birds will rebound through the positive cycles of their own population growth. Then the human population will switch back to the positive phase of the oscillation until it is dampened again.

Positive feedback can lead to exponential, even system-disrupting growth. The concept of such "runaway" feedback is much with us these days, in everything from climatic greenhouse effects to cascading economic trends and our understanding of the geological histories of Mars or Venus. In local biological networks, runaway positive feedback can bring about dramatic outcomes if it is not braked—think of that flu or strep or of an algal bloom in a pond where runoff fertilizer phosphates have lifted constraints on nutrients.

Effects of positive feedback

In evolutionary dynamics the possible consequences of positive feedback have only begun to be explored, but it is clear that they can be dramatic and unexpected. In the 1990s geologist Douglas S. Robertson and biologist Michael C. Grant offered several scenarios (Robertson 1991; Robertson and Grant 1996). They started by using a positive-feedback model to explore the mysterious, much-observed tendency of individual species to evolve toward larger body size (a tendency known as Cope's Rule). Here is how their model worked: In any particular species, given loose genetic determinants of body dimensions, there will be a range of body sizes, with mid-range sizes more frequent than outliers, such that the distribution, if graphed, would form a bell curve. The peak of the curve, the most frequent body size, represents also a fitness peak, the selected body size that confers greatest fitness for the species in its environment, perhaps because it enables the greatest hunting prowess or escape from hunters, perhaps because it facilitates the best diffusion of resources through the system, or for many other reasons. At the same time, an *intra*-species fitness pressure might also exist, conferring some advantage on slightly larger individuals—for example, in an animal population, an advantage in competing for mates. The second dynamic would then exert a positive feedback pushing the size distribution of the whole population away from the greatest overall fitness. The average body size of the species would grow larger, and its peak on the size distribution graph would shift, no longer coinciding with that of the fitness graph.

In addition to offering an explanation for Cope's Rule, this model has implications concerning the evolutionary consequences of positive feedback. It suggests that sometimes the feedback can lead toward a breaking apart of a single population, with (in graphic terms) ideal sizes now clustering in two peaks, each corresponding to a *different* fitness peak determined by different criteria—a mechanism that might eventually result in division of the populations and even full speciation. More dramatically, as the feedback militating for larger bodies worked *against* overall fitness it might, in an extreme form, lead a species to shift far from its ideal fitness, toward a fitness level compromised enough to bring about its extinction—selection for *un*fitness! Projected onto a large, multispecies canvas, Robertson and Grant suggested, such a mechanism might explain the regular, oscillatory rhythm of extinctions in groups of species (that is, whole *genera*) that paleontologists have observed and wondered about.

The mechanism also suggests how a species could last for long periods in a slowly changing form, with negative and positive feedback more or less in balance, only then to reach a threshold where positive feedback became dominant and the species underwent quick transformation. This would conform to Stephen Jay Gould's "punctuated equilibrium" model of phylogenetic history, in which long periods of evolutionary stasis are interrupted by dramatic change. This kind of dynamic might even explain quick bursts of evolutionary diversification, such as that preserved in the Cambrian deposits of the Burgess Shale (Gould 1989). But at the same time, Robertson and Grant showed, an opposed dynamic *decreasing* diversification could be another possible outcome of positive feedback. This could occur if its nonlinear, exponential effects overwhelmed other forces or constraints at work in a system. In this case the feedback would "lock in" certain evolutionary and morphological tendencies, enabling them to dominate the selective landscape and in effect pointing natural selection in one direction rather than another. Such locking might explain the evolutionary history that seems to have followed the Cambrian explosion of the Burgess Shale period, which witnessed a rapid narrowing of the diversity of floral and faunal body types.

Robertson and Grant make a strong case for the pervasiveness of feedback in the evolution of life. They go so far as to see in it a fourth element in the Darwinian algorithm: inheritance, variation, selection, and feedback. This seems to me a categorical misstep, for we need to understand feedback not as an additional element so much as the systemic *relation* of the first two elements as they unfold in time and across generations to bring about the third. Misstep or not, however, their assertion is no exaggeration of the paramount role itself of feedback. Compound interest may not be the most powerful force in the universe, but in the history of life its effects have been large.

Feedforward There is one more refinement that needs to be introduced here, although it is not usually included in discussions of evolutionary feedback. Perhaps this is because it is not technically part of a feedback dynamic at all but defines an element acting on that dynamic from the outside.

With inevitable simplification, we can think of biological feedback systems as closed, insofar as the feedback itself concerns the specific traits that form the loop. The linked selective histories of clover and bee or peacock and peahen, for example, form closed cycles. In the first case, ease of access to nectar, through traits selected in both species, closes the loop; in the second, the female's registering of a male display as a sign of a worthy mate does the same. There are of course other factors involved, which open these closed loops to the outside world. In the case of the peacock and peahen I have specified one of these in the differential predation on peacocks and conceptualized this as the enchaining of one feedback loop (predator and prey) with another (male and female birds). But to think of the selective history whereby peacocks developed elaborate tail feathers and peahens came to admire them as a closed system is not only analytically productive but also an accurate historical description as far as it goes.

Outside any biological cycle, however, there will always be other elements that determine or control it *but are not in any significant way altered by it*. They form the conditions for a feedback loop but are unaltered by the feedback. An obvious example for most biological systems is the climate. Sudden fluctuations in Upper Pleistocene global temperatures changed human ecosystems but were effectively untouched by humans' changed behaviors in response to them. The climate changes stood outside the alterations brought about by all the feedback cycles of which humans formed a part. Other obvious examples of such external elements are astronomical cycles, tectonic shifts, and volcanism. Three million years ago, the joining of the North and South American continents by the volcanic emergence of the Isthmus of Panama resulted in the so-called Great American Interchange of countless species. It altered all their cyclic relations with their ecosystems and drove many of them to extinction—notably the dominant predators of South America, the flightless "terror" birds. But the volcanism itself was unchanged by those alterations, not part of the feedback cycles it altered.

In control systems theory, elements sending a signal into a system from outside it are called *feedforward* elements. These external elements have the capacity to determine and alter in radical ways the systems they interact with. Defining these elements, however, will in many instances not be an absolute, black-and-white affair but a question of relative scales of systems and controls, and this is especially true in biological situations. The global climate, for example, might at first thought seem always to be a feedforward element. In considering local, small ecosystems this is for all practical purposes true; the impact of the ecosystem of a pond on the world's climatic system is insignificant. But the example today of anthropogenic climate change is a clear instance of a single species remaking the global climate, in effect ushering what had been a feedforward element into the workings of its own feedback cycles (Tomlinson 2017); and this is not the first instance of such biotic remaking of the global climate. We can imagine scales between the pond ecosystem and human climate disruption where it will be difficult to judge whether an element plays a feedforward role, untouched by local feedback dynamics, or enters into those dynamics.

An additional complexity of feedforward elements will play a central role in the model of human evolution I offer later. In certain circumstances, feedback cycles can generate elements that come to stand outside them with emergent, organized dynamics of their own. These can then function *as if* they were controlling, feedforward elements, altering and determining the systems from which they arose with little change to themselves. In the broad view, these can hardly be thought of as pure instances of feedforward, since they took shape historically out of the very feedback dynamics they now seem to control; perhaps "pseudo-feedforward" or "quasi-feedforward" is a more accurate designation. But at a later historical moment, when their autonomous organization has taken full form, their operation can effectively mimic the impact of a true feedforward control.

We will see that hominin cultures came at a certain point to generate an abundance of such feedforward controls; these I call *epicycles*.

My definition here of the concept of feedforward is the one sanctioned in classic control theory, but it is a definition often misunderstood. In discussions of biological cycles and evolution in particular, the phrase "feed forward" is habitually used as a synonym for positive feedback, the "forward" thus distinguishing it from negative feedback, and it is even sometimes used to connote a simple causal chain, A→B →C, where there is no feedback at all. The distinction of feedforward from positive feedback or causal chains without feedback is, however, a categorical one, and the confusion is important to avoid.

CHAPTER 3

THE CONSEQUENCE OF FEEDBACK II: NICHE CONSTRUCTION AND CULTURE

:1:

Niche construction

In the instances of coevolutionary feedback examined so far, the environment figured mainly in vague references to "resources" and in external, feedforward elements such as the climate. But another major initiative in the extended synthesis brings the environment to the heart of the feedback relations of evolution.

The customary role of the environment in Darwinian evolution is that of a static determinant of selective pressures. It is a court of appeals for organisms, sitting in implacable judgment as phenotypic variants in a population approach its bench for their fitness to be assessed. Its own changeable nature asserts itself occasionally—usually slowly, as when the local or global climate shifts or geological subsidence makes a shallow sea deeper, but sometimes abruptly, as when a volcano erupts or a meteor strikes. Abrupt changes result in quick reshaping of selective pressures and extinction rates increase, sometimes to the point where paleontologists mark them off as mass extinctions. But for the most part the judging environment is slow to change.

This view of the environment and its role in natural selection is not wrong, especially in broad focus, where it is akin to the idea that certain aspects of the environment stand outside evolutionary feedback cycles in their capacity to operate as feedforward elements. And it has a pragmatic research value: by leaving a set of factors out of the equations of population genetics, it has simplified the models and facilitated much analysis. But feedforward independence does not characterize every aspect of an organism's environment, and the conventional view, if not wrong, is incomplete. There is no organism so miraculously light-footed as to have no impact on its environment. A beaver building a dam changes the physical, hydrological profile of its ecosystem, altering the local conditions for countless other species. A colony of bacteria shifts the balance of resources in its microenvironment, depleting what it takes in as food and adding waste products. Such changes wrought by organisms, accruing over many generations, can have large cumulative impacts. They can alter even global environments. The earliest photosynthetic bacteria arose more than two billion years ago in an atmosphere with no free oxygen, and their metabolic release of oxygen eventually brought the composition of the earth's atmosphere close to the one that sustains most life today. This so-called Great Oxygenation Event is an instance of nonhuman, biotic, worldwide environmental change.

Feedback and environment In all such changes, from the local instance of a pond created by a beaver dam to the global one of the earth's atmosphere remade by photosynthesis, organisms engineer their ecosystems (Jones, Lawton, and Sachak 1994, 1997). They *construct their niches*. As they do so, the environment comes into a different kind of relation with them, for the niche, once constructed, exerts an altered force back on the constructors. Across generations, as organisms change their environments in ways determined by their phenotypes, they redirect the selection acting on their own future progeny, including whatever varying traits might arise among them. The earliest castorids, ancestors of modern beavers, lacked tree-gnawing incisors, broad tails for swimming, and (presumably) a suite of instinctual dam-building behaviors. These emerged in tandem, gradually and incrementally, as the castorids took more and more to the water and as environmental changes brought about by their incipient, lodge-making behaviors shifted the selective pressures on them. The environment they altered fed back into the selection acting on them. Two billion years earlier, the buildup of oxygen in the atmosphere altered selective forces, eventually changing the nature of the bacteria that had caused it (along with many other microorganisms: the Great Oxygenation Event is also known as the Oxygen Catastrophe, since it is thought to have led to a mass extinction of anaerobic organisms that had evolved up to that time).

These examples describe feedback cycles unfolding across evolutionary time between organisms and their partly constructed environments. The cycles are not sporadic and exceptional; they are inevitable and pervasive. If the force of natural selection on a population or species is a matter of constraints and affordances in its environment, then (in the Darwinian model) it will create a feedback loop within the reproducing, varying population. The consistent environmental force will favor one variant over another. But if (in the extended model) organisms alter their environments and the balance of affordances and constraints in them, then the feedback loop embraces both organism and environment. Organisms and environment alike are shifting conditions, mutually bound, in the dynamics of selection.

In their important book *Niche Construction: The Neglected Process in Evolution* (2003), John Odling-Smee, Kevin Laland, and Marcus Feldman conceived of this effect in tripartite terms. There are (1) the traits of organisms that alter the environment, (2) the environmental resources that are altered, and (3) the traits of organisms that are differentially selected by virtue of the changed environment. The traits, 1 and 3, are linked or mediated through the resources, 2. A couple of general possibilities arise from this schema of niche-constructive feedback. If the traits producing the altered environment (1) and the traits selected by it (3) are in the same population of organisms—in the same species, say—then, across generations, the species will pass down within its lineage a legacy of environmental change (and the changed vectors of selection that come with it). Odling-Smee and his coauthors refer to this legacy as an "ecological inheritance" and regard it as tantamount to a second line of inheritance, rivaling in importance genetic inheritance. They think of it in informational terms: genetic information is sent via phenotypic traits (no. 1 above), transmitted through the middle term of the environment altered by the traits (2), and received by other traits (3) and the genes behind them. As ancestral beavers changed their environments in ways that put into play, incipiently, the selective advantages of aquatic lifestyle, traits furthering that lifestyle—broad tails for swimming, for example—were preferentially selected by aspects of the changed environment. The genes supporting those favored traits in effect gained greater fitness. The watery habitat constructed by making lodges was increasingly bequeathed down the lineage as an advantageous one, and selection changed accordingly.

Coevolution expanded

There is a second general possibility, because the environmental changes (the ecological inheritance) will affect not only the species that created them. Imagine that the first, environment-altering traits (1) are in one species (for example, oxygen production in a species of bacteria) and that the second traits, affected by the altered environment (3), are in another (other microbes, the metabolisms of which benefit from greater oxygen concentration). Then we have a coevolutionary situation extending between species, but one in which we have now recognized a role for environmental mediation. This underlines the close connection between coevolution and niche construction. In chapter 2 I depicted the predator-prey dynamic in a schematic fashion, as a relation of two species only and their specific traits for attack and defense. It is clear, however, that there is always more to the story than hunter and hunted: the food of the prey species, the other resources both species use for their habitats, the water they both require—not to mention the many additional species, from microorganisms to competitors for food or habitat, with which they interact. Both species are nodes in a web of Van Valen's (1973) competing optima that extends in all directions around them; they do not exist merely in an engagement with one another. And these many optima are defined not only by living things but also by inorganic aspects of the environment—by *abiota* as well as biota.

If we look hard at any coevolutionary dynamic, we will see that a one-on-one view of it is a simplification, for it always needs to be conceived as ecologically engaged across a broad range. It needs to be understood as altering everything in that range as it is altered by it, and so as involving niche-constructive feedback. Another way of thinking about this convergence of interspecies feedback and niche construction is to realize that the niche of any organism is an array of elements that includes not merely the abiota around it but all the other organisms with which it interacts.

Environmental information Let us examine a bit more closely the flow of information that Odling-Smee and his coauthors identify in the feedback cycles of niche construction. This information is genetic in nature for most organisms; that is, the traits in the cycle (1 and 3) are the phenotypic expression of genes controlling them. (The qualification "most" makes an exception for cultural animals, for which the information is both genetic and cultural; I return to these below.) So the whole cycle through which the information flows involves genes as well as traits and the environment:

DIAGRAM 1

Genetic information	→	Phenotypic trait	→	Environmental alteration	→	Phenotypic trait	→	Genetic information

The genes and the phenotypic traits that reflect them are carriers of information; this is an easy, if informal notion. But the affected aspects of the environment, *living or nonliving*, are also information carriers—a more challenging idea. We can think of this environmental information for now as the conformation of some aspect of the environment to the genetic and phenotypic prerogatives that act on it. The beaver builds a dam, channeling genetic information through its phenotypic behavior; the hydrology of the local ecosystem, remade by the dam, conforms to the information in the beaver's genotype and also conveys this conformity to other organisms by acting on their phenotypes. Then, if the pattern of the altered ecosystem is persistent enough through generations for a selective bias to form in the affected populations, the frequencies of genes expressed in the affected elements of their phenotypes will be altered.

The information that started in the beaver genotype has been cycled through the material environment, biotic and abiotic, and has come around to alter the selection acting on other genes in subsequent generations of beavers (or other organisms, since the traits and genes on the right-hand side of diagram 1 need not belong to the same species as those on the left-hand side). The genes at the beginning and end of the cycle are connected by the passage of information through the environment. They form a cycle binding genes, traits, and extraorganismal aspects of the engineered niche.

The notion of information implied here will need further development in chapter 4. Odling-Smee, Laland, and Feldman burden it with several qualifiers that I find unhelpful, calling it "semantic" information, writing of its "meaning," and seeing it as constituting "knowledge," if usually of an "entirely noncognitive" kind (2003, 180–83). We will need to sort through this tangle of concepts in order to make important distinctions concerning different varieties of niche construction and different kinds of organismal action in the world, and this in turn will help us better to understand the actions of cultural animals. For now, however, we can move forward with an unqualified notion of information.

:2:

Modeling niche construction

It is one thing to suggest a feedback cycle between a few generations of beavers and their local ecosystem, quite another to propose that this cycle carries long-term evolutionary consequences that shift the dynamics of natural selection and thereby change the genetic makeup of a population. But this second proposal stands at the heart of *Niche Construction*, and to bolster it Odling-Smee, Laland, and Feldman (2003, chap. 3 and related appendices) set out to model the consequences mathematically.

They employ the basic tool of population genetics, *recursion equations*, with which changes in the relative frequencies of genes can be projected statistically across generations. The simplest recursions plot two variants, or *alleles*, of a single gene against one another. By making certain starting assumptions (for example, the beginning frequencies of each variant in a population or the nature of reproduction of the variants) and then building factors into the equation to reflect pressures selecting one allele or the other, equations can be constructed that project future frequencies of the two, including points at which they reach equilibrium—that is, points when their frequencies stop changing from generation to generation.

Odling-Smee and his coauthors make these recursion equations more complicated by building into them two genes, each with two alleles. This is called a "two-locus" model, since it tracks the frequencies of genes at two loci on the chromosome in relation to one another. For their purposes, they take one of the genes to be the one expressed in the trait that alters the environment (they label it E, for its environmental impact), while the other gene is the one that is differentially selected by the altered environment (A). We can add these genes to the feedback cycle diagrammed above, using their letters and also distinguishing the two alleles in (or variants of) each gene with italic upper- and lowercase:

DIAGRAM 2

Genetic information	→	Phenotypic trait	→	Environmental alteration	→	Phenotypic trait	→	Genetic information
Gene E: alleles *Ee*				Resource R				Gene A: alleles *Aa*

In the middle of the diagram is another addition: a particular environmental resource R. This is the crucial novelty the authors introduce in their two-locus model that enables it to represent niche construction. The representation depends on two guiding assumptions. The first is that the two alleles of E, *E* and *e*, differently control a trait of the organism that determines different conditions of resource R in the environment (for example, the *E* variant might increase it and the *e* deplete it). The second is that the fitness of the gene at the end of the chart differs for its two alleles, *A* and *a*, in ways that vary with the condition of R in the environment (increase in R might select for *A*, depletion of it for *a*). Selection of *A* or *a* is now "frequency dependent" on R; R is in turn frequency dependent on the E alleles.

We can readily see that the two-locus model, which without R plots differential selection (different fitnesses) of A alleles as a function of frequencies of E alleles, does something more with the addition of R: it plots differential selection of A alleles *as a function of the condition of an environmental resource determined by* E alleles. The connection of genes through environmental mediation has been captured in the model. Along with it, at least rudimentarily, niche construction as a selective force has also been captured.

The equations elaborating this model can include several additional measures: coefficients representing the fitness of different combinations of alleles, independent of niche construction; other coefficients that measure the sensitivity of the fitness of A to different values of R; and others still that reflect the relative strengths of these fixed and frequency-dependent fitness measures. When different values are entered into these equations and the recursions are run—yielding a model of tendencies across generations—the results give strong support to the hypothesis that the feedback channeled through the environmental resources altered by E will ultimately change the frequencies of A alleles in a population. The representation of niche-constructive feedback in the model affirms its capacity to alter natural selection.

But the model achieves more than this, since it can capture a number of other real-world complications. Odling-Smee et al. build into some of their equations variables for other, independent selection pressures, not caused by niche construction, so that in some of their calculations E is the only determinant of selection, while in others forces external to the ERA causal path are also involved. These external selective forces can act on either E or A and can add to or oppose the force of niche construction. (Perhaps *E* makes the organism more vulnerable to a parasite than *e* does, for example, or *A* strengthens defenses against a predator more than *a*.)

Implications of the model

We begin to see how complicated even a simple model can quickly become, and the results of all this calculation are correspondingly varied and complex. Assuming that allele *E* (but not *e*) changes R in a way that favors selection of allele *A* (not *a*), then a situation without external selection in which *E* is active—"pure" niche construction—will lead to the spread of *A* through the population (its "fixation" in the population). If external forces are at work selecting against *E*, then *e* will become fixed, but either *a* or *A* will spread through the population, depending on the relative weighting of the external forces and the decrease or accumulation of R. External selection favoring *E* can lead to fixation of either *A* or *a*, depending again on the relative weighting and on changes in R.

External selective forces acting on the A locus, meanwhile, result in still-greater complexity. If the selective force generated by niche construction redoubles this external force, the situation is clear, and *A* is fixed. But if niche construction opposes the external selection of *A*, it can, depending on its strength, override the external force, leading to fixation of *a*; the idea implicit here that niche construction can effectively counteract other, independent selective tendencies is suggestive. Or niche construction can lead to a stable mixture of *A* and *a* in the population—a "polymorphic" equilibrium of two different alleles (hence two different traits).

Odling-Smee and his coauthors also consider the way selective dynamics change according to differing temporal impacts of E on the environment. If the impact of E on R changes the environment quickly, then the effect on the frequency of A alleles might also be quick; there can be a short-term temporal linkage between changes in E-allele and in A-allele frequencies. (Their example is a spider, with E a genetic locus affecting web building, and A a locus affecting defense of the web.) If, instead, the impact of E is to accumulate R gradually across generations, then the linkage might be weakened, and a time lag of some or many generations might open between the changes of E and A frequencies. (Their example is an earthworm, with E affecting its processing of the soil, resulting in gradual changes in its composition, and A affecting a phenotypic accommodation to soil conditions.) Given the other selective forces that might be involved, this time lag can be shortened, or, if those forces oppose niche construction, the allele favored by it can become rare or even disappear from the population before it is spread by the niche-constructive effect. In this case the time lag might introduce a selective inertia into the population, resulting in no response (or a sluggish one) to the changes wrought by E.

Simplicity of the model Notwithstanding its evident complexities, it is clear that the model of niche construction offered by Odling-Smee, Laland, and Feldman has its limitations. These stem especially from the reduced dimensions the model encompasses (two genes, two alleles, one resource) and the straightforward causal assumptions built into it (this gene affects this behavior, which changes this resource, etc.). The authors are as aware of these limitations as any of their readers, and they name other aspects that might be built into a more elaborate model: the size and density of the population of constructors, the growth rates and life histories of its individuals, and the scale of alteration of the ecosystem (2003, 134, 164–65). But they are at the same time aware that the bluntness of the model is its strength. It is what enables the model to represent selective dynamics in a way that affirms the fundamental niche-constructive hypothesis: that organisms' remaking of their environments can have a legacy effect on the course of their own and other organisms' natural selection.

The virtue of simplicity in this model bespeaks a more general truth about such biological models. It is because of their simplified assumptions and low dimensionality that they can capture essential features of far more complex processes of the real world and represent their consequences—predicting their courses in a general way. Moreover, the simplicity is a virtue made of necessity for, as we have glimpsed, however drastically the models reduce real-world complexity, they nevertheless burgeon quickly into complication of design and difficulty of calculation. Their dimensions, in fact, must remain few in order for them to be tractable. Two other intrepid modelers of evolutionary process have affirmed this need for simplicity, since "to substitute an ill-understood model of the world for the ill-understood world is not progress" (Boyd and Richerson 1985, 25).

The problem with modeling of this sort, then, is never its simplicity. When problems arise, they stem from overreaching extrapolations from the model back to the world it so sparsely samples. What the model of Odling-Smee, Laland, and Feldman shows is that niche construction can alter natural selection, and can do so in various ways, depending on other factors involved in it. To apply this lesson to human evolution, as we will do in later chapters, requires specific cases, empirical evidence from them, and nuanced deep-historical reconstruction.

: 3 :

Culture and evolution

At the end of the presentation of their formal model, Odling-Smee and his colleagues drop a small bombshell: "There is no requirement for niche construction to be directly determined by a gene at one locus before it can alter the selection of a gene at a second locus. Niche construction can depend on learning . . . ; the consequences for a gene at the A locus would still be the same, provided that the effect of the niche construction on the environmental resource was similar for learned and unlearned behavior" (2003, 166). They understand full well the import of this statement and spend much time investigating it later in their book: animal learning and social learning and social learning passed on through generations, which we took in chapter 1 to form a good preliminary definition of *culture*—all these can alter the historical course of natural selection. Odling-Smee, Laland, and Feldman introduce cultural processes into the selection-altering feedback of niche construction. Doing so, they introduce culture into the machinery of evolution.

There is a long and troubled backstory to this move. The idea that learned behavior can change natural lineages reaches back before Darwin in the form of the discredited Lamarckian notion that "acquired characteristics" can be transmitted to future generations. How, some evolutionists ask, does the niche constructionists' case differ from Lamarck's giraffe, which stretched its neck to reach higher leaves and so passed a longer neck to its progeny? A clear explanation of the differences between old-style and new-style "acquired traits" is needed, and also a mechanism whereby acquired traits could alter the gene pool.

There is another worry, of an ethical and political sort. Linking natural selection and human culture was the foremost strategy of the late nineteenth-century popularization of Darwin's natural selection. It took root as social Darwinism, the application to cultural development of the "struggle for existence" driving natural selection. And the implications of this kind of thinking turned darker still in the first half of the twentieth century, when the emergence of genetics fostered the fantasies and practices of eugenics. Given its large genetic component, the modern evolutionary synthesis has always needed to separate itself from projects of human genetic control, and these issues have not become less pressing in the twenty-first century.

Unsurprisingly in this context, most evolutionary biologists in the years following World War II had little to say about culture or social learning. This situation began to change, however, by the 1960s, and from the 1970s on two new fields addressed the question head-on: sociobiology and evolutionary psychology.

Sociobiology Sociobiologists proposed that social organization in animals originated from the selection of advantageous behaviors and its shaping of genotypes, and they set to work using the tools of population genetics to model how this happened (Wilson [1975] 2000; Alcock 2001). Their proposal of an adaptive origin for sociality was in a general way unimpeachable, of course: viewed from sufficient distance, animal behaviors are all traceable, more or less directly, to natural selection. Moreover, in asking how adaptive behavior could result in the observed complexities of animal sociality, sociobiologists helped to focus attention on some knotty puzzles in the modern synthesis: the evolution of group cooperation and altruism, for example, or the division of populations into reproducing and nonreproducing castes, as in certain insect societies (it is called *eusociality*, and Darwin had already pondered it at length in the *Origin*). Exploring such issues led to advances in thinking about the level at which natural selection operates. It had usually been taken to be the level of individual organisms contributing their genes to gene pools (or not), but now it was seen that selection could act on whole populations as well.

Despite these contributions, sociobiologists came under attack for the stripped-down adaptationist program that characterized much of their work. In their view, the patterns of sociality and even of human culture took shape according to the adaptive advantages they conferred. Selection on them was a direct reflection of their capacity to enable their practitioners to get their genes into the pool. The voices raised against this view, and in favor of a more complex interplay of nature and nurture than the sociobiologists envisaged, were many (Sahlins 1977; Gould 1981; Lewontin, Rose, and Kamin 1984).

Evolutionary psychology If complexity of the nature-nurture interplay was the rule for ant colonies and elephant herds, it had to be so a fortiori for human societies. So the debate over adaptationism and culture reached a fevered pitch when sociobiology gave rise to the field of evolutionary psychology, which aimed to study the evolution of modern "human nature" and the human mind. Here the adaptationist program of sociobiology was extended to diverse features of human culture, tracing them to prehistoric humans' advantageous responses to the exigencies of their lives (Barkow, Cosmides, and Tooby 1992). The proposition proved seductive, and books by the dozens have emerged since the 1990s proposing the ancient adaptive advantages gained by cultural practices of all sorts and at many levels of specificity—books aiming to explain everything from cannibalism to board games, from altruism to infanticide, and from storytelling to religion. The popular outgrowths of this kind of thinking have promulgated Paleo-diets and interpreted Jane Austen's plots through dynamics of sexual selection. (The last move comes in the area self-styled "literary Darwinism"; see Kramnick 2011.)

Such tracing of specific modern behaviors to Paleolithic selective pressures derives its appeal partly from its generality. And the generality is in fact unanswerable, since it cannot be doubted (for example) that Austen's plots have *something* to do with mating and sexual selection, even if the question remains open of what we gain by noting this in deepened understanding either of Austen's novels or of sexual selection in human societies. But generality has its dangers, since it breeds a kind of explanatory bluntness in which behavioral determinism is often endorsed (as when sexual aggression in men is ascribed to genetic predestination), the effects of culture are stinted, and the nuances of specific histories, deep or otherwise, are overlooked.

These tendencies plague the central project of evolutionary psychologists: understanding the emergence of the modern mind. The idea that it is the product of its evolution is another hypothesis that can hardly be gainsaid, but in itself it explains no more than connecting Austen's plots to mating does. To fill in the details, many evolutionary psychologists offer a schematic and intuitive evolutionary sketch, commonsensical but only loosely tethered to specific cases and evidence. They propose that adaptive pressures organized the mind into cognitive *modules*, rich in content that enabled us to face the challenges of the Pleistocene world. Particularly important among these modules are a physics module (triggered, for example, to evade a falling rock or throw a spear), a biology module (distinguishing what we might eat from what might eat us), a psychology module (recognizing similar minds around us), and an "agency detection device" setting off alarms at every leaf stirred in the breeze (Barkow, Cosmides, and Tooby 1992; Barrett 2004; Tremlin 2006). These were advantageous for our ancient forebears and were called into operation in specific lived situations; and in similar situations today, the argument goes, they operate still.

This kind of reasoning has, once more, a kernel of truth. We are undoubtedly armed with a panoply of basic capacities, genetically determined, that enable us to confront and make sense of the world. But the idea of an adapted, modular mind can be extended very far, without much close examination of human prehistory. The most influential modularist of the 1990s, Steven Pinker, used modules to propound a Chomskyan "language instinct" as well as to tell us in general "how the mind works" (Pinker 1994, 1997)—claims that took little account of evolutionary forces and that moved a leading proponent of modularity from the 1980s, Jerry Fodor, to protest the oversimplification and respond with a tract acidly entitled *The Mind Doesn't Work That Way* (Fodor 2000, 1983).

Even when deep history was taken into account, the generality of modular theories encouraged explanations of a reductive cast. The best-developed case in point comes from a leading Paleolithic archaeologist, Steven Mithen. In *The Prehistory of the Mind: The Cognitive Origins of Art and Science* (1996) Mithen traced the emergence of the modern human mind to an alteration of modular structure, in which what had once been unconnected, multiple intelligences (technical, natural historical, and social; compare the physics, biology, and psychology modules) were joined in a fluid cognition. The unconnected modules of Neandertals explain for Mithen all the puzzling differences of these humans from us, while the breakdown of the borders between the modules and the resulting cognitive flexibility enabled the "big bang" of *Homo sapiens*' culture, located at the Middle/Upper Paleolithic border. And the primary stimulus that brought about the new fluidity was modern language, which created a metamodular reflection on and integration of the previously disjoined intelligences (Mithen 1996). Mithen's account has wielded considerable influence in its merger of modular evolutionary psychology with the specifics of Paleolithic archaeology. But at the same time his view of Neandertal limitations and differences from us has been challenged by much new evidence. His portrayal of the Middle to Upper Paleolithic explosion grows less convincing with each new season's gathering of archaeological evidence. Furthermore, his linguocentrism, which makes language the decisive stimulus of modernity, is an oversimplification of a more complex set of developments (Tomlinson 2015). I will return to these issues in chapters 6 and 7. For now it is enough to note in *The Prehistory of the Mind* both the generalizing strengths and the blunt-force weaknesses of evolutionary psychology.

We can draw two lessons from the debate over sociobiology and evolutionary psychology. First, the idea that our minds are adapted phenomena, while unexceptionable, will not readily yield tools precise enough to discern the interactions of culture and biology in recent human evolution. Mapping these details requires a careful weighing of deep-historical evidence about ancient cultures alongside general evolutionary models. Second, any hypotheses of causes for adaptations in the hominin clade over the last three million years that see them to involve biology alone and do not make room for hominin culture will necessarily be partial hypotheses, since there is no moment in this history when noncultural selective forces could have acted in sheer isolation from culture. Whatever the interaction of biology and culture may be, it has been at work since the beginning of the hominin lineage, long before such recent variants as *Homo sapiens* or *neanderthalensis*. This does not mean that the interaction did not change during the coalescence of human modernity, and it is this change that I aim to describe in the coming chapters. In the most general terms, in fact, the challenge of describing the coalescence is to understand how a mechanism that generated *all* biological forms could finally produce *some* forms that channeled information in ways that could rebound on the system from an *ideational* place. The cyclic rebound is an aspect of what we wish to understand throughout hominin history, and here again we come around to a familiar feedback pattern, now involving culture as well as genes. Genes are always important; but in cultural animals, so is culture, and to try to track the one without the other is folly.

: 4 :

Dual-inheritance theory

The general relation between biological and cultural evolution that sociobiologists and evolutionary psychologists posited gave clear priority to biology and natural selection. They sometimes devoted ample attention to mechanisms of cultural evolution (Wilson 1978; Barkow, Cosmides, and Tooby 1992; Alcock 2001), advancing hypotheses about the nature of the dependency of cultural development on biological adaptation, but there was less consideration of a cyclical relation by which cultural evolution fed back into and changed biological phylogenies, and the consequence of feedback figured little in the equation.

This one-way relation characterized also a whole stream of accounts of cultural evolution, starting in the 1970s, that ran alongside the debates on sociobiology and evolutionary psychology. This literature proposed that cultural selection occurs in ways *analogous* to natural selection, and it dug deep into the mechanisms shared between the two, using mathematical methods devised to analyze changing patterns in populations of genes. Culture was understood as a repertory of replicators similar as a whole to a genotype. The repertory formed an inheritance system distinct from genetic inheritance but developing through parallel processes of variation and selection; for this reason the formalizing of it came to be known as *dual-inheritance theory*.

Several researchers stand out in this initiative. From the side of population genetics and biological modeling came two teams: Luigi Luca Cavalli-Sforza and Marcus Feldman (more recently, a coauthor of *Niche Construction*); and Robert Boyd and Peter Richerson (from whom I took my baseline definition of culture in chapter 1). They began their work in the 1970s, and major studies soon emerged: *Cultural Transmission and Evolution: A Quantitative Approach* by Cavalli-Sforza and Feldman (1981) and *Culture and the Evolutionary Process* by Boyd and Richerson (1985). Meanwhile, sociobiology inspired cultural anthropologists to weigh in on evolutionary issues, often but not always in hostile fashion (Sahlins 1977); and in 1991 anthropologist William Durham merged the earlier work by the biologists with ethnographic interpretations to offer a general theory of cultural and biological selection in a book entitled *Coevolution: Genes, Culture, and Human Diversity.*

The biological strain

The researchers on the biological side were all trained in the statistical analysis of populations and their change, and they worked to extend the recursion equations of population genetics to cultural transmission (Boyd and Richerson 1985). They schematized and quantified the ways cultural knowledge was passed through three kinds of transmission: vertical (parent → child), oblique (other adult → child), and horizontal (peers within a generation). They specified analogies between biology and culture, likening "errors" in the transmission of cultural knowledge to genetic mutations and random perturbations of what knowledge was transmitted to genetic "drift," significant but nonselective alterations in gene frequencies in small populations. They analyzed forces that could shift the balances of cultural knowledge across generations, changing a population's "cultural phenotype." These included forces whereby each generation might slightly alter knowledge gained from its parent generation through reactions to its own environment, thus "guiding" the variation of culture in adaptive ways. And they included several vectors of bias that might arise in transmission: direct preference for one variant over another based on its inherent features; frequency-dependent, or "conformist," biases, where frequent variants are favored more than infrequent ones; and indirect preferences brought about because of the association of one cultural trait with another. All these forces were figured into equations that quantified the cultural content itself along several lines: either-or ("dichotomized") choices, alternatives that could be blended in different mixtures, and alternatives chosen from among a set of proximate options.

Today the limitations of this modeling of cultural selection are clear enough, especially in the schematic view of culture it imposes. As in the case of niche construction, the researchers had to reduce the complexities of cultural transmission to simple categories (vertical vs. oblique, either-or vs. blendable, etc.) in order to make the calculations practicable; this is part and parcel of the modeling process. But in their quantification of such central issues as learning and social transmission they also relied on analyses borrowed from the "hard" social sciences—quantitative sociology, psychometrics, rational choice psychology, and the like. Along with this influence came definitions that today seem limiting and arbitrary, such as the restriction of cultural replicators to ideational patterns only, excluding practical behaviors, or the attributing of behavior to self-conscious decision-making. As approaches to human culture in either its ethnographic or its historical dimensions, such analyses have not held up well. Despite its limitations, however, the biological strain of cultural selectionism offered an impressive mathematical model of cultural evolution operating through the Darwinian forces of inheritance, variation, and selection. It affirmed the possibility of a cultural selection analogous and parallel to natural selection and pointed to many complex patterns that could result from it.

The anthropological strain

Durham's analysis (1991) of the parallels between cultural and natural selection came, in the same years as the mathematical modeling, from the "soft" social science of interpretive anthropology. Durham adopted the term *coevolution* for these parallels—confusingly, since the word, as we have seen, had usually connoted feedback mutualism between the coevolving entities, and there is little of this in his model. He saw the two inheritance systems, cultural and genetic, as connected in some aspects, disconnected in others. The connection comes about because in each system traits of varying fitness are replicated, varied, and differentially selected, and so the two systems share the same large-scale, Darwinian mechanism. On the cultural side, Durham proposed ideational dynamics by which replicated traits are transmitted but at the same time transformed across generations, thus providing the cultural grist necessary for the selective mill to operate. This transformative transmission arises because the selection of cultural traits is a product of both "primary," genetically determined values and "secondary" values shaped by societal knowledge, traditions, and histories. Durham argued for the clear predominance of secondary values; these are the driving force of cultural selection, shadowed by broader, genetic determinations but also buffering its operation from them.

The weak point of Durham's account is, unsurprisingly, his view of a cultural trait. Durham advances a conceptual, ideational definition of traits and distinguishes them sharply from behaviors. Settling on Dawkins's meme as the term of art for these ideational units, he extends the terminology to further the genetic analogy: "allomemes," named by analogy to alleles, are the different variants of a meme that carry different fitnesses and are differentially selected. The difficulties that plague memetics arise here too. How are we to draw the borders Durham's memes require to make them the discrete ideational units he envisages? And, if we cannot define these borders, how do they function as varying replicators? And what, we could ask more generally, does replication mean in a fluid, nonparticulate situation? The usefulness of memes, as we saw in chapter 2, breaks down precisely because they artificially check the proliferating meanings—and indeed the levels at which meanings proliferate—of cultural gestures that might seem identical. This difficulty spins out of control even as Durham tries to specify the range of competing allomemes, which, he writes, could be "differing techniques or strategies for procuring subsistence resources; alternative schools or sects of religious thought coexistent within a population; differing conceptions about the length of a postpartum sex taboo; or varying definitions of a word or label" (1991, 189–90). An allomeme could, then, take the form of a single flint-knapping technique helpful in procuring food or of a whole lithic industry. It could be Methodism or all of Islam, an articulated ideology on female impurity or a communal rule of sexual abstinence followed by rote. All these allomemes, moreover, are transmitted in Durham's model by decision-making, rational actors. Here the implausibility of fixed memes weighs heavily, and the reduction of cultural complexity necessary to model them has become extreme.

Gene-culture coevolution

Whether taking their start from biology or anthropology, cultural selectionists emphasized analogies: the similar but independent designs of the mechanisms of cultural and biological evolution. They also considered the *functional* relations of natural and cultural selection, where the mechanisms mesh, but here their thinking tended to remain close to that of the sociobiologists, who posited a one-way causality with little sign of feedback from culture to biology. Natural selection, in this view, provides a large, guiding context for cultural selection and shapes its course; along the way, cultural selection might also assert its quasi-independent force. Durham worked hard, using especially his distinction of primary and secondary values, to articulate a plausible meshing of genetic and cultural gears that could register the genotypic impact on cultural phenotypes while preserving a degree of autonomy for culture. Boyd and Richerson, for their part, modeled situations in the interaction of cultural and natural selection in which the cultural trajectory could detach itself from and even counteract the genetic one.

One consequence of this thinking about the situational autonomy of cultural from natural selection was that both anthropologists and biologists began to appreciate the need for a more detailed historical understanding of how their meshing came about. A shift toward deep-historical evidence is apparent, for example, in the papers Boyd and Richerson presented in the 1990s, where Pleistocene humans sometimes figure importantly and titles like "How Microevolutionary Processes Give Rise to History" suggest that a more humanistic, rounded, and fluid approach to culture is starting to take hold (see Boyd and Richerson 2005). In the same years, Boyd and Richerson came to understand that the functional relation of genes and culture could not be adequately described as a one-way causal operation, no matter how complex. They saw that the meshing of the two inheritance systems had to result in changed conditions not only for cultural but also for natural selection. By 2005, in a new and important book, they could write: "Once cultural traditions create novel environments, environments that can affect the fitness of alternative *genetically* transmitted variants, genes and culture are joined in a coevolutionary dance" (Richerson and Boyd 2005, 190). This dance, unlike the gene-culture relation of their earlier work or Durham's or Cavalli-Sforza and Feldman's work, was now envisaged as a two-way interaction involving feedback. *Gene-culture coevolution* was the name applied to the new, updated, feedback version of cultural selectionism. The keyword *coevolution*, co-opted by Durham for a nonfeedback parallelism, reverted to the cyclic implications of its classical usage. Now, however, instead of marking the mutualism of two species, it applied to the interaction of the two inheritance systems, genes and culture.

We can see from the inclusion of the *environment* in the quotation from Boyd and Richerson that genes and culture are not the only ingredients of this new view. Culture changes environments, leading to changes in the fitness of genes: the link of genes and culture runs through an environmental detour and externalization already familiar to us. We return to niche construction.

: 5 :

Cultural niche construction

Odling-Smee, Laland, and Feldman (2003, chaps. 6 and 9) had also recognized this coevolutionary dance. "Niche construction," as they wrote, "can depend on learning." To understand the role of social learning and culture they adapted their models in a deceptively simple way, substituting cultural actions or behaviors for the E gene, whose alleles, in the noncultural model, had controlled differing traits that changed the environment. We can represent their new model with an adjustment in the flowchart of diagram 2, removing the E alleles and replacing them with cultural behaviors shifted to the phenotypic position:

DIAGRAM 3

[Genetic information]	→ Phenotypic trait	→ Environmental alteration	→ Phenotypic trait	→ Genetic information
	Cultural behaviors *E* and *e*	Resource R		Gene A: alleles *Aa*

What does this adjustment accomplish? First, we see that it unites all three inheritance systems that we have by now named: the genetic and cultural systems of the cultural selectionists as well as the ecological inheritance of the niche constructionists. *E* and *e* are assumed to be variants of cultural behaviors in a population, carried across generations by social learning. The cultural variants differently affect resource R, altering the environment, and this change in turn differently affects the fitness of another varying phenotypic trait and the genes behind it, selecting preferentially for allele *A* or *a*. Cultural, environmental, and genetic legacies are bound together in the cycle, and information flows along cultural as well as genetic and ecological channels.

This step added something new to the work of sociobiologists and cultural selectionists. Odling-Smee and his coauthors fully understood the older, sociobiological models, which indicated that cultural practices could make it more likely that some animals in populations would reproduce than others, thereby changing the gene pool of future generations. But if this was the only kind of effect of culture on evolution, the only connection of culture and natural selection, then only culture's direct effect on survival value mattered—and this was just the kind of hard-core adaptationist program that had elicited objections in the 1970s and 1980s. Odling-Smee et al. saw also that culture could evolve through a parallel selective track running alongside natural selection. This had been the starting point for the dual-inheritance theorists, who posited analogous mechanisms for the two, with biology usually in the driver's seat. Now, in addition to these two possibilities, the niche constructionists offered a third. The feedback of the altered environment on the organisms whose culture alters it creates a new kind of mediation between culture and genes.

Somewhere behind the cultural behaviors in this model, of course, stand genetic determinants of them. In the case of advanced animal cultures, however, these are usually not tight controls, and indeed they can be of the broadest sort. In order for animals to have social learning and culture at all, for example, they must have certain genetically determined capacities. But here is the crucial point: however tight or loose the genetic control, it is the *differences* between the cultural traits at work (between *E* and *e*) that count, and these are not genetically determined but produced as variants in traditions of social learning. For this reason I have bracketed the genes at the left-hand side of diagram 3; they can be safely left out of initial calculations of culture's effects in the niche-constructive cycle.

A new role for acquired traits

Since the cultural traits are acquired characteristics of the animals that deploy them, linking them to inherited changes might seem to raise the specter of a discredited Lamarckism. But this objection is easily discounted, since the acquired traits posited to have an impact on selective forces acting on later generations, their genes included, are set off in a distinct, nongenetic inheritance track. The difference between the old, fallacious Lamarckism and cultural niche construction is a question of different information moving along different channels and the distinct legacies thus formed. Lamarckism proposed the transmission from a parent giraffe to its young of a neck stretched through use and habit, which involved an impossible crossing of channels between a behavior—the effort of the parent to reach higher leaves—and a biological inheritance. Cultural niche construction proposes, to outline a case in point, that novel toolmaking techniques 500,000 years ago—acquired traits—could so alter the lived world of the hominins who invented them and passed them along that selective forces were altered, finally altering the genetic makeup of their population. We may or may not wish to call this Neo-Lamarckism, but, whatever we call it, the principle is clear. The lithic techniques in the example form information that runs always in a nongenetic, cultural channel, and their impact on the genetic channel is mediated by the inheritance of an altered environment. There is no need for a toolmaking gene to appear anywhere in this evolutionary history.

Mathematical modeling

The mathematical models corresponding to diagram 3 show the necessary simplification and inevitable bluntness that we saw already in the models of noncultural niche construction. Here the bluntness concerns culture and recalls the cultural selectionists' schematic modeling of it. The model's distinct units of culture, *E* and *e*, are analogous to the alleles labeled with the same letters in diagram 2; memes are not mentioned, but they might as well be. As with the noncultural models, however, complexities are quick to gather. Odling-Smee and his team quantify the intergenerational transmission of the cultural alleles using cultural selectionists' techniques. They formalize vertical transmission (parent → offspring) in three ways: unbiased, in which offspring show cultural practices in the same proportion as their parents do; biased, in which offspring of mixed parental cultures preferentially select one practice or the other; and incomplete, in which offspring can sometimes not choose the practices of their parents. There are other complicating factors also, which run parallel to those of the noncultural niche construction equations. External selective forces, for example, can be worked into both sides of the equation. That is, different behaviors (E and e) or different A alleles or both can be preferentially selected by forces not connected to the niche construction.

Effects of cultural niche construction When all the math is said and done, two general results stand out (Odling-Smee, Laland, and Feldman 2003, chap. 6 and appendix 4). First, the model indicates that cultural niche construction can be a major force driving selection and changing gene frequencies in future generations. Second, it suggests that the dynamic by which it does so is anything but a simple one. The equations show no straightforward linear or proportional correlation between the frequency of cultural practices and that of genes. This kind of correlation often emerges from the calculations of the cultural selectionists, which do not include the environment in the middle, but the interposing of mediating resources in the niche-constructive model expands the range of possible outcomes.

Counterselection and buffering The model represents the force of cultural niche construction in three aspects: effects of strength, tempo, and variety. Niche construction can, in the right conditions, effectively oppose other, non-niche-constructive selection, even if that is strong. In this case the cultural effects can spread or even fix an A allele that is otherwise selected against. As Odling-Smee and his team see, this "counterselection" is very suggestive for animals of high cultural attainment, especially hominins, since it indicates that their culture can effectively "buffer" them against natural selective forces. This buffering effect is not as obvious as it might seem. Applied to ancient humans, it is inadequate to think of it as a mundane case of lighting a fire at the mouth of a cave to stay warm in an Ice Age winter, even though this act buffered the humans involved against the inadequacies of their biological makeup for the niche they inhabited. The buffering needs to be imagined instead on a broader scale: as a tradition of cultural uses of fire (see Wrangham 2009), extended over many generations, that effectively shaped the developing genome of a population which, without fire, would have built other niches, lived out other selective histories, and ended up in a different genotypic space. The challenge of extrapolating the effects of culture is often not met in qualitative thinking about hominin evolution. The recursion equations of cultural niche construction address this challenge in mathematical terms, and deep-historical descriptions must follow suit.

Acceleration

Also suggestive for hominins is the indication that cultural niche construction can accelerate slower tendencies of natural selection. This can be a simple function of selective vectors in both areas that point in the same direction and redouble each other's force, speeding the spread through a population of alleles preferentially selected—the opposite of counterselection, in other words. But it can also manifest a more basic difference between the tempos of cultural and genetic change in a population. Odling-Smee and his colleagues presume that cultural states will change faster than genetic states, and quantitative analysis of such hard-to-measure rates suggests that this will usually be the case (Perreault 2012). We can presume, then, that a new and effective cultural technique in a band of hominins will spread through the group more quickly than most selected genes will do. An isolated population of hominins might gain control of fire and begin singeing their foodstuffs, especially meat, in effect "predigesting" them, and thus derive more nutrition than otherwise. These behaviors might be adopted quickly by the whole group and then, over generations of practice, accelerate a cascade of changes in the oral tract and digestive system that had already begun for other selective reasons.

Another effect of the rapid pace of cultural change involves time lags between effects on R (environmental resources) and the resulting effects on A (genetic alleles). The noncultural model indicated that these could be long, but the cultural model shows that the rapid pace of the spread of a cultural trait (E or e) can reduce the length of time lags, as its effect on R can be correspondingly rapid. This will not be true for all cultural effects, however. Human depletion of megafauna populations through hunting in the Late Pleistocene is an instance of cultural niche construction, likely repeated independently in many locales by separate populations, and it probably had dramatic impacts on many different local environments. But these impacts, although pronounced, in most cases built up gradually—more slowly, even, than some noncultural changes in gene frequencies might occur. So the effect of cultural traits on natural selective forces acting on human populations might involve long as well as short time lags.

Culture-induced thresholds

The variety that arises in the model can result in situations where the fixing of A or a is a touch-and-go affair, with one balance of cultural traits and environmental resources selecting A, and another, slightly different one selecting a. When external selective forces are taken into account, this can lead to threshold conditions where the selection switches directions, to oscillating patterns where it flips back and forth, and to stable polymorphic equilibria of *A* and *a*. This complexity appears especially when cultural transmission in the model is biased (progeny preferentially choosing between cultural alternatives) or incomplete (progeny not always taking on their parents' ways). For example, biased transmission, in changing the frequencies of the cultural variants, could result in a switch from preferential selection of one A allele to the other. Strong cultural bias might bring about the switch in a short time, while weak bias might do so slowly or even take so long that the allele "switched to" was rare by the time it came to be selected. In populations divided from one another, Odling-Smee et al. suggest, this could lead toward fixed genetic distinctions and ultimately speciation—driven by a cultural dynamic! And all these complicated patterns are generated by a simplified model. In the more complex and fluid conditions of real cultural transmission among hominins—with oblique and horizontal as well as vertical transmission and less particulate and distinct cultural states at play, to mention only the most obvious complications—we might expect to find frequent shifts, switch-points, oscillations, and polymorphic populations.

Closing the feedback loop Odling-Smee and his colleagues devote little attention to the aftereffects of the changes in genetic frequencies brought about by cultural niche construction—the aftereffects, that is, of the shifted balance of *A* and *a* at the right-hand side of diagram 3. But they appreciate that these in their turn could provide an altered genetic, biological foundation for further cultural elaboration, even sponsoring whole new cultural capacities. As a possible case in point in hominin evolution they offer the so-called "expensive tissue hypothesis" of Leslie Aiello and Peter Wheeler (1995). This is one of many hypotheses that have been advanced to help explain the evolutionary increase in brain size among ancient hominins. Aiello and Wheeler propose that meat eating (a dietary shift dependent on technology, hence cultural) allowed for a reduction in the size of the metabolically costly gut organs (a selected, genetic event). Reduction in these organs eased constraints on selection for expansion of the *other* most costly organ, the brain (additional selected, genetic events), and this expansion underpinned further cultural elaboration. The feedback cycle here runs from (1) cultural interaction with the environment to (2) new genotypes altering certain traits, then to (3) a cascade of further genetic changes, and finally to (4) additional elaboration of culture, as the cycle is closed. In the schema of diagram 3, such a model in effect removes the brackets I placed around the gene at the left-hand side and draws an arrow connecting the right-hand gene back to the beginning:

DIAGRAM 4

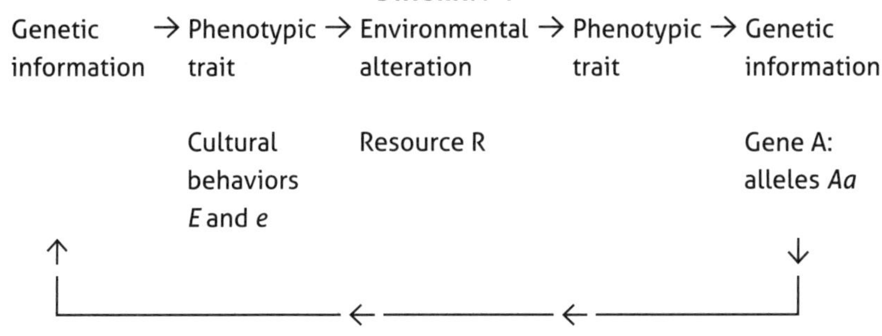

Once we see that culture can be a force in the natural selective history of a species, we must conclude that the resulting new genotypes could likewise exert selective forces back on the capacities underlying its developing culture. Any genotypic change that resulted in new phenotypic traits would have the potential to do so, after all. Indeed, recent modeling of cultural niche construction has suggested that such linkages could be very powerful, fixing both cultural practices and gene-based traits even when the practices are costly and the traits disadvantageous. There is good reason to think that this kind of cycle occurred often in hominin evolution (Rendell, Fogarty, and Laland 2011). For human ancestors, the "coevolutionary dance" of culture, environment, and genes was an unending round.

: 6 :

The place of feedback

The approaches surveyed here track the deepening incorporation of culture into conceptions of natural selection and biological evolution across the last few decades of thought about evolution. By now, culture has found an important position in the extended evolutionary synthesis, and it is widely understood that its patterns of interaction with genes and environment bring about reciprocal dynamics that are usefully gathered under the rubric of feedback.

The feedback cycles described here and in chapter 2 concern several distinct sets of elements. The classic coevolutionary situation involves the mutual interaction of two species, while the sexual selection feedback cycle takes place within a single species. Niche construction, instead, draws populations of organisms and the environments they partly construct into the cycle, with feedback running from the population through ecological changes and back to the population. For cultural animals, gene-culture coevolution highlights the effects of genes on culture or culture on genes, which must always involve a middle term, the phenotypic expression of genes. Cultural niche construction, finally, interposes the environment in the midst of these genetic, phenotypic, and cultural elements. In all cases, the feedback exerts its force across generations, shaping and directing the operation of selection. The many faces of feedback we have seen, finally, are all dynamics in the evolutionary change of populations through the selection of inherited features. That is why they are all extensions of the modern evolutionary synthesis and not departures from it.

These models are much simplified—necessarily so, we can once more, and for the last time, affirm. The notion of a feedback cycle will be useful only as long as we keep in mind its inherent *incompleteness*: that is, the fact that we extract its elements for heuristic purposes from real-world systems that involve whole networks of mutualisms, constellations of connected parts whose feedback relations engage many other parts on multiple levels. A straightforward instance, mentioned earlier, of such multiple engagement is the case of the enchained, intra- and interspecies feedback loops involved in selection for the peacock's tail. The incompleteness of discerned feedback cycles is everywhere apparent in the models of gene-culture coevolution and cultural niche construction. Moreover, the control of feedback loops in large or small measure by feedforward elements external to them forms a categorical proposition of their incompleteness, and of course, viewed in broad enough perspective, feedback cycles in the living world will *always* be controlled by feedforward dynamics. Explanatory richness will be found by including local controls and scrutinizing the details of finer perspectives.

Culture The views of culture in these models carry their own, heavy burden of incompleteness and simplification. In order to understand the dynamics of hominin evolution, and particularly late hominin evolution, we will need to consider the results of such models in the light of an elaborated conception of culture and its emergence. Forming this will be the work of chapters 4 and 5. This conception will be a qualitative, not a quantitative, model of culture. It will be pieced together from reasoning about foundational requirements for culture joined with some deep-historical evidence of the practices and capacities of our ancient ancestors and their relatives—evidence of a sort that will feature even more in later chapters. Finally, my model will be built, as far as is possible in a retrospective venture such as history, according to a prospective vantage: ancient hominin culture viewed, that is, from the bottom up.

CHAPTER 4

HOMININ CULTURE FROM THE BOTTOM UP

:1:

Human niche construction

As we have seen, cultural selectionists, who from the 1970s on explored the analogies of cultural and biological evolution, posited two parallel inheritance systems, one genetic and one cultural: the dual-inheritance theory. As they developed their understanding of the functional links between the two systems, they saw that it was not enough to track a one-way impact of genetics on culture, for if the social transmission of acquired knowledge alters the fitness of its bearers differentially, its feedback loops must interlink with those of natural selection. With this reasoning culture was ushered into the heart of the machinery of evolution, and the study of gene-culture coevolution was born.

Like body and soul in the old Cartesian dualism, however, cultural and genetic inheritance seemed unable to touch one another. What was the middle term, the medium through which cultural evolution could make its force felt in genetic selection? Enter the niche constructionists, who proposed a third distinct inheritance system, an ecological inheritance connecting genetic and cultural legacies. The environment reshaped by living things was the mediating term. It did not matter whether its reshaping was accomplished through noncultural phenotypic traits or the "cultural phenotype"; in either case, the impact of a rebuilt environment could rebound in the same way on natural selective pressures. In the full picture of biocultural evolution offered by the niche constructionists, none of the three separate inheritance systems is independent of the others. Instead, they are enchained one with another, loop upon loop, in a network of feedback interactions.

In the particular case of hominins, however, the picture requires something more. The key difference in our own biocultural evolution is a hypertrophied system of cultural inheritance. Already by 1.5 million years ago this had probably outstripped the cultural inheritances of any other lineages the earth had seen. By half a million years ago cultural achievements were such that we begin to recognize in them characteristic patterns we can call "human" as readily as "hominin." And by 100,000 years ago, in some human populations, culture had begun a runaway acceleration that I will aim to explain in later chapters.

As paleoanthropologists and ethologists know, and as most evolutionists will readily concede, this hyperdevelopment of culture sets hominins apart. What they will not so readily grant is that it changed the qualitative nature of the feedback interactions of genes, environment, and culture. This is nevertheless the argument I will advance: that the cultural development of human populations reached a level of complexity that created a new *kind* of interaction in the dynamics of biocultural evolution. This is the general change in late hominin evolution that the gene-culture coevolutionists and the niche constructionists have missed. It is not taken into account in the quantitative models of cultural evolution that paleodemographers, taking their cue from the evolutionists, have offered, and it is absent from the rich and plentiful descriptive accounts that archaeologists and paleoanthropologists have provided. It is the piece missing from our accounts of the appearance of modern humans, and it explains how this event qualifies as a major transition in the history of life.

Information To explore hominins' hypertrophied culture requires careful definition of some keywords, for only by building from this conceptual foundation can its special systems and novel powers be revealed. The keywords have to do with what is transmitted in culture, with the nature of its transmission, and with its relation to a much broader spectrum of animal activities in the world. The word we start with is one that niche constructionists already made their own in an informal usage: *information*.

For Odling-Smee, Laland, and Feldman (2003), all niche construction involves the transmission of information. The manner in which it does so distinguishes three channels of information, which in turn define the three inheritance systems—genetic, ecological, and cultural. Information from genes is carried in phenotypic traits of all organisms; in their reshaping of the environment, it moves through a material, ecological medium; and in cultural animals it runs also along a separate track that intersects with the external niche and, through it, with genes. In all three varieties it is the current that flows through the feedback circuits. But what exactly is this information? Is it something more than a metaphor for the connections posited by niche construction theory?

Launching the Information Age in 1948, Claude Shannon defined information as a problem of communication, that is, of "reproducing at one point either exactly or approximately a message selected at another point"

(Shannon and Weaver 1949, 31). Shannon's famous schematic diagram of communication showed an information source, a transmitter, a receiver, and a destination for the message—and in the middle a channel of a certain capacity and with inevitable distortion, or noise. Information is any signal or signals drawn from a finite set of possibilities and communicated along the channel. It can be quantified—this was Shannon's decisive insight—as the measure of the probability that the message at the receiver will correspond with that at the transmitter. Quantities of information are related to the probability that the succession of units in the original message will be received just as it was sent. The greater the number of possible successions, as determined by the size of the lexicon of units from which those in the message are selected and the length of the message, the greater the information conveyed in maintaining the succession across the channel. Information in this conception bears a close relation to the physical concept of entropy, since the correspondence between transmitted and received signals it quantifies is in effect a measure of distance from randomness, that is, from maximum entropy. Information, in sum, is correspondence achieved in the face of randomness.

Of the many features that follow from this conception of information, two are particularly important here. First, information is essentially reduplicative in nature. It is a conformation between entities at two points, and so it arises only in a doubled existence. The measurement of information in units—for example, the *bits* Shannon chose, named after the binary logarithmic scale he used—easily suggests that information is a single, isolatable thing, like the gravitational force of the moon. But it is not; instead, its measurement captures a relation between things, a correspondence located at a gaugeable distance from randomness. The moon exerts a gravitational pull on the earth, and the earth on the moon; the information flowing between moon and earth is the probability of far-from-random, predictable effects that the correspondence brings about on each, for example, the tides.

Correspondence versus content

This notion of the moon and earth locked in an informational correspondence may seem surprising, and it brings us to the second, somewhat counterintuitive feature of information in the wake of Shannon. Since information is a correspondence, it need not involve a content or meaning transmitted. "Semantic aspects of communication are irrelevant to the engineering problem" is the way Shannon put it (Shannon and Weaver 1949, 31), but it is not just our thinking about engineering that his conception altered. The correspondence view of information entails a relation, but it does not necessarily involve the specific kind of relation that creates content, that is, the asymmetric presentation of one thing in another, or *re*presentation. Information transmitted may carry with it the reference, meaning, or "aboutness" that arises in such representation, but if so, this is an add-on, inessential to its nature. Aboutness or meaning, as we will see, is essential only to certain types of information: signs.

Of course, the information humans communicate is habitually about things; we are a species that constantly, constitutively loads phenomena with meaning. This, along with the fact that the word *information* long predated Shannon in an informal usage that concerned meaning and content, is why the idea of contentless information feels counterintuitive. It is also why most theories of information since Shannon, from his colleague Warren Weaver and cotheorist Donald MacKay down to anthropologists Paul Kockelman and Terrence Deacon, have aimed to illuminate the relation of information and meaning (Shannon and Weaver 1949; MacKay 1969; Deacon 2007–8; Kockelman 2013b). This aim is marked also in ideas of information in evolutionary biology. John Maynard Smith, for example, conjures meaning into the genetic code by definitional fiat. Meaning for him is simply the ability of genes or proteins to function "in a way that favors the survival of the organism"; since genetic information specifies adapted phenotypic form and function, the relation of genotype and phenotype is inherently meaningful and "semantic," even carrying "intentionality" or aboutness (Maynard Smith 2000, 179). Genes, in short, are about the proteins they produce and vice versa.

There is, however, a short circuit at work in this assigning of meaning to biological information. By conflating a correspondence view of information with a referential or meaning-laden one, it keeps us from comprehending information in its largest, most bracing sense, as a nonrandom correspondence of things in the cosmos causing predictable and nonrandom effects. The conflation muddles important distinctions, and the curious consequences that can result from it are displayed in Deacon's (2007–8) analysis. His starting point in defining information is the second law of thermodynamics, which requires that the lessening of entropy involved in all informational correspondence must involve work and energy. Deacon views this expenditure as extrinsic to the "information bearing process" itself, an "external perturbation" that forms a part of its context or environment. The nature of the (intrinsic) information process, however, gives us clues as to the nature of the (extrinsic) expenditure, and it is these clues, tantamount to so many pointers, that Deacon considers to be the origin of reference or aboutness. The relation between the intrinsic process and the perturbation—the work decreasing its randomness—thus gives rise to a "second-order" kind of information, "referential information" (2007–8, 139).

The slippage in Deacon's reasoning concerns the relation between his intrinsic "process" and extrinsic "work." This is nothing other than a correspondence of the sort that creates information in the first place, shifted to a different level and extending between entities different from those corresponding in his intrinsic process. The relation of process and work is, in other words, simply more information. Deacon has discovered reference in the connection of two informational correspondences, one of which he labels "work" rather than information, but there are no grounds for seeing in his second correspondence a categorical difference from the first, or for thinking that the linkage of two informational correspondences can of itself yield reference.

We can illustrate the case with an example, the relation between a bacterium and its environment. There are many informational correspondences at work within the cell and many additional, distinct ones at work between it and its environment. Following Deacon's reasoning, the first would be intrinsic processes, and the second, extrinsic perturbations of the first—work operating on them. But this is a strange "bacteriocentric" view of the world, in which the cell membrane separating inside from outside has the power to differentiate categorically what are in reality no more than multiple informational correspondences. Viewed from the perspective of organism + environment rather than of organism alone, these processes are all categorically similar elements in the informational network of a system poised far from equilibrium.

Extending reference as far as Deacon does—in effect, locating it in the informational correspondence itself—raises the additional question of how it is registered or observed in the world. Who or what reads his clues? Deacon's term for this registering is *interpretation*, a very human-oriented word, and in fact most of his examples of interpretation are human ones—detectives at crime scenes, scientific experimenters, or everyday language users—in which the interpretive agency is clear. But these can help us little in fathoming the interpretation that goes on between a bacterium and its environment, and so we are left, in his nonhuman cases, with uses of the word that cannot be more than metaphorical. Thus, the entropy level of nucleotide sequences is "interpreted to" the amino acids to which they correspond (2007–8, 144), inherited genetic information is "interpreted" (here Deacon puts the word in scare quotes) in the structures of the inheriting organism (180), and trees interpret or misinterpret their local environments (187–88).

Causal covariance

An altogether less problematic idea of information, and one close to Shannon's conception—to "Shannon information," as it is called—is offered by Jerry Fodor. He emphasizes the reduplication built into all information and the distinction between it and meaning; indeed, for him meaning not only is irrelevant but has come to be the key term of contrast: "There is a lot less meaning around than there is information. That's because all you need for information is reliable causal covariance.... Information is ubiquitous but not robust; meaning is robust but not ubiquitous" (Fodor 1990, 93). *Reliable causal covariance* is a powerful phrase that captures the correspondence at the heart of information; its implication of causally bound entities also retains, in a general form, Shannon's transmitter and receiver. As a postulate of minimal requirements, the phrase sets the bar very low, pervading the cosmos with information in all manner of simple and complex systems. Information is not restricted to living things and the biosphere but ubiquitous far beyond them (as my example of moon and earth implies). Thinking of information along these lines has in recent decades helped biologists to understand the relation of genetic and epigenetic factors in development, the place of living systems in broader, nonliving contexts, and even the origin of life (Sterelny 2000; Kauffman 2000; Deacon 2012b).

At the same time, Shannon's and Fodor's conceptions also mark the foundational place of information *within* the biosphere. Though we may balk at the idea of DNA "interpreted to" amino acids, we cannot think of the protein-producing correspondences of DNA, RNA, and amino acids without thinking of reliable causal covariance and hence information. (It is significant that the code-obsessed World War II years that saw the development of Shannon's theory of communication also witnessed the first definition of life in informational terms, by physicist Erwin Schrödinger, a decade before the structure of DNA was determined; see Schrödinger [1944] 1992.) In these molecular correspondences there is, again, no content at stake. To speak of information "in" a DNA molecule, "about" the surrounding cell, its metabolism, its development, or its reproduction, is to speak loosely. A DNA molecule is a two-stranded sequence of nucleotide bases in a certain spatial arrangement; there is nothing "in" it at all, in informational terms. Information arises from its conformation with other molecules, according to a covariance that is reliable in causing proteins to be produced. In a similar way there is no information about a galaxy in the physical structure of a black hole at its center, but there is information aplenty running between them. While, however, the examples of black hole and galaxy or earth and moon are similar to that of DNA and proteins, they are not identical in informational terms. This is because living examples of information involve the networked, complex products of selected, evolved histories. Because of this historicity, the constraints on the probabilities of informational correspondence in living things arise, are organized, and persist in different ways than in informational systems outside life (Maynard Smith 2000; Deacon 2007–8, 2012b). In both cases, nevertheless, content or aboutness is irrelevant to the conformational effects brought about.

What is true of the molecular networks in organisms is true of organisms as well. Since all living things involve correspondence and covariance of highly elaborated and reliable sorts, they are all *information processors*. In conversation some years ago I asked evolutionist Stuart Kauffman for a biologist's definition of information, and he paused only briefly before responding: "A bacterium swimming up a glucose gradient." It was not a definition, but an adroit example. To say that information is *not* being processed by the bacterium in a covariance with its environment, or indeed that it is not being processed in the bacterium's internal metabolic and genetic networks, would defeat any useful conception of information. Information in the form of systemic chemical correspondences constantly cycles inside the cell and between it and its surroundings, as the cell maintains itself far from equilibrium through causal correspondences of myriad sorts. But still we have not crossed over to a world of content or aboutness, and to imagine that our bacterium forms or transmits meanings makes little sense—except informally, on the part of the observing scientist in the habit of thinking of related things as "about" one another. The bacterium does not think about glucose and where to find it; there is only causal correspondence or its lack, in a continuum of possible measures forming Kauffman's gradient.

Information in niche construction

Shannon information describes the information that niche constructionists assign to their model, at least insofar as this concerns their first two inheritance systems, genetic and ecological. This information is the contentless correspondence that connects genes, actions in the world, and the environment. Covariances—or, better, networked assemblages of covariances, products of evolved histories—conform the material world to genes as their phenotypic expressions reshape it. A beaver dam, a spider web, or an algal bloom will, to some degree of reliability, have certain kinds of effects on a local ecology and not others; the degree of reliability is itself the measure of information. In all this niche-constructive correlation, however, neither genotype nor ecosystem is about anything.

Culture enacted by animals, the third category of niche construction theory, introduces something new. It is a special type of phenotypic expression that, in its niche-constructive impact, layers aboutness, content, and meaning on top of information—robustly so, as Fodor might say. Culture is always about something. It is never mere Shannon information, because it is made by a small group of animals whose interaction with the world creates content and proliferates meaning. This reasoning appears to be circular: culture has content because cultural animals make content. To break the circle, we must examine those special animal interactions with the world. Though they are not always cultural, from them all culture springs; because they are replete with content, so is it.

: 2 :

Sign and semiosis Aboutness is not a trait of information, but it is fundamental to culture. The question of how it arose in the world takes us to my second keyword and draws an important distinction between it and the first. The new keyword is *sign*, and the distinction divides a cosmos filled with information from a much smaller realm within it of signs. This *semiotic* realm is full of content and meaning because signs are always signs *of*: indicators of objects or effects other than themselves. Signs form the foundation in the world of what philosophers call *intentionality*, the power of perceiving entities (philosophers usually say "minds") to be directed at something, "to be about, to represent, or to stand for, things, properties and states of affairs" (Jacob 2014). Following Fodor, this meaningfulness of signs suggests that they are not nearly as widespread as information; if all living things are necessarily information processors, only some of them are *sign makers*. Within this group, inside the semiotic realm, is an even smaller category of culture makers. The first implication of this hierarchy is clear: information is logically and ontologically prior to sign, as sign is to culture.

A further distinction of terms is useful here, between *sign* and *signal*. The informal usage in the life sciences of "signal" usually does not imply a semiotic process, and no sign is necessarily involved when a biologist writes of the "signaling connections" between genes or of "signals" to an organism "from the physical world" or of "the way external signals become incorporated" in organismal reactions (Lewontin 2000, 11, 61, 64). More often, in such informal usage, "signal" refers to nothing more than a trigger or threshold of causal efficacy involved in the correspondences that define information. It is what makes operative the covariance at the ends of the information channel, an aspect of information processing that is "about" nothing.

These distinctions and the limited arena I describe for the sign will seem a challenge to many researchers in *biosemiotics*, a field that came of age in the late twentieth century as the study of extrahuman semiosis or sign making. Biosemioticians today discover signs everywhere in living things: in all manner of animals (cf. zoosemiotics), in plants (phytosemiotics), in bacteria and other microorganisms, and even in the molecular networks and interactions basic to all of these (Kull 1999, 2000; El-Hani, Queiroz, and Emmeche 2009; Favereau 2010; Emmeche and Kull 2011); for Kalevi Kull, a leading proponent of this position, "it has become widely accepted within biosemiotics that the semiotic approach is an appropriate tool to describe all living systems, down to the first cells" (2009, 12). Even the conformations of DNA to RNA and the alignments of nucleotide codons to amino acids are seen as instances of sign making. In the light of the definition of information offered above, however, it will be apparent that this extension of the sign is tantamount to equating it with information. The equation is unwarranted, and to understand why this is so—to see, that is, what is *added* in the correspondence between sign and object that makes semiosis different from information and its signals—we need to pause over the theory that stands behind both the biosemioticians' position and my own: the semiotics of Charles Sanders Peirce (1839–1914).

Understanding Peirce is not easy. In a long, iconoclastic career he wrote incessantly, leaving scholars to wrestle with a *Nachlass* of something like 100,000 manuscript pages, within which are proposed many typologies of signs, multiple terminologies, and shifting definitions of even his central concepts. Toward the end of his career he tallied sixty-six sign types, distinguished in ways that semioticians still labor to explain (Short 2007; Nöth 2011; Atkin 2013). Nevertheless, Peirce's views form a semiotics that has influenced many fields and remains compelling, especially because he focused on the *process* of signification more than on the *structure* of the sign. There is nothing inert, nothing static about Peirce's semiosis. It is an active relation between perceivers and the connections they perceive of one thing to another. And, though Peirce himself usually thought of his perceivers in human terms, his processual orientation gives his semiotics a broad, extrahuman application (Peirce himself recognized this late in his career). It is this breadth that has allowed his ideas to serve as a basis for biosemiotic work and that will found my own stance developed below.

The Peircean interpretant

What sets the semiotic process apart, in Peirce's view, is a condition in it—it was also, for Peirce, a basic, quasi-Kantian category of our experience—that he called *thirdness* (Peirce 1955, chap. 6). Peirce realized that signification involves not merely the relation of a sign to its object. This, if it could be isolated, would amount to an instance of *secondness*, a relation rather like the correspondence that defines information. And, regressing further, we might consider either the sign or its object, standing alone, as instances of *firstness*. But these are nonexistent constructs; firstness has no signifying power, since a sign without an object cannot function as a sign, and secondness likewise cannot signify, since there is nothing that its sign-object link would signify *to*. Semiosis requires also a third, mediating element that intrudes upon the sign-object connection, *a relation to its relation* struck up in the perceiving thing that comes to be aware of the connection (Peirce 1958, 387–90; Kockelman 2013a). This element is arguably Peirce's signal contribution to the theory of the sign; he called it *interpretant*. It is the completing element, Peirce explained, in a triadic complex of relations that enables signification: "A *Sign* . . . is a First which stands in such a genuine triadic relation to a Second, called its *Object*, as to be capable of determining a Third, called its *Interpretant*, to assume the same triadic relation to its Object in which it stands itself" (quoted from Nöth 2011, 457).

Peirce's interpretant is central to understanding the emergence in some living things of capacities beyond information processing alone. We can think of it, to start, as an effect of the sign and its object on the perceiver. "The Sign," Peirce wrote, "creates something in the Mind of the Interpreter, which something, in that it has been so created by the Sign, has been, in a mediate and *relative* way, also created by the Object of the Sign. . . . And this creature of the Sign is called the Interpretant" (1998, 2:493). The interpretant comes close to the *meaning* of the sign that arises in the sign's relation to its object, and sometimes Peirce refers to the interpretant as meaning. But "interpretant" also connotes something more active in the perceiver, and this ingredient is crucial. The interpretant is like a stimulus and a response: a calling the world issues to the perceiver to interpret and also the perceiver's reaction that forms the bond connecting sign and object. These become sign and object *only* in their connection brought about by the interpretant, which completes their relation-to-a-relation, or thirdness. Semiosis is never a passive registering of ready-made connections plucked from the world; instead, it is an interaction with a world of things that *might* be connected, and the interpretant names this constructive, poietic interaction. Anthropologist Kockelman puts the matter this way: "For something to constitute a sign, it must be able to not just stand out in an environment (be a difference) but also be sensible to an organism. And for something to constitute an interpretant, it must be able to not just stand up in an environment (make a difference) but also be instigated by an organism" (2017, 27).

The interpretant, then, names the *intentional* aspect of a sign, its directedness toward a meaning, content, or object, and at the same time it specifies something *attentional* that brings the sign into existence. It is a salience (Rumbaugh et al. 2007) that rises up in the reactions of certain complex organisms to stimuli, folding the environment so as to bring into relation two things in it and, in doing so, make one of them about the other. What is this attention that leads to semiosis? It seems to be connected to animal capacities such as memory and conditioned or associative learning. It is not ubiquitous in living things; plants and many animals seem not to possess it, and bacteria and other microorganisms surely do not. Organisms such as these make their way in the world through information alone, albeit abundant and complex information in even the simplest cases: information, again, that has no necessary concomitant of aboutness in it but that is nevertheless unlike information in a galaxy because it is the product of a selected, evolutionary history. (The fascinating recent discoveries about plants' interactions with their environments are comprehensible in terms of such evolved information and require no phytosemiosis to explain them; a careful distinction of sign from information would help to resolve the debates waged today over the nature of plant capacities.) Attention, finally, comes down to a focusing of broader capacities of perception that some animals manage, by which certain aspects of environmental stimuli are singled out as important or relevant in a situation, while others are ignored (Tomasello et al. 2005).

Through their attentions, semiotic organisms build on top of the informational foundation of all life a new dimension of experience in which one thing stands in for another, a surrogate presenting it to perception in its absence: a *re*presentation. This is a word notorious in cognitive and related studies for the haziness and multiplicity of its meanings; the semiotic approach can help to bring it into clearer focus. Representation appeared in the course of evolutionary history along with semiotic organisms that introduced a new kind of repetition into the world. On top of a repetition by *dis*placement—informational correspondence at the ends of Shannon's channel—there appeared repetition by *re*placement. There is every reason to think that the advent of organisms capable of this reflects a major transition or set of transitions in the history of life.

Hierarchic signs

In one of his many analyses of the interpretant, Peirce distinguished in it three degrees, nested like Russian dolls: "In all cases," he wrote, the interpretant "includes feelings; for there must, at least, be a sense of comprehending the meaning of the sign. If it includes more than mere feeling, it must evoke some kind of effort. It may include something besides, which, for the present, may be vaguely called 'thought.' I term these three kinds of interpretant the 'emotional,' the 'energetic,' and the 'logical' interpretant" (1998, 2:409). The typology captures the interpreting activity Peirce discovered at the heart of semiosis, and it also signals another basic feature of complex types of signs: their hierarchic and recursive nature. A sign that comes about through an emotional interpretant alone involves no other type, but any sign involving an energetic interpretant will involve an emotional one as well, and any sign involving a logical interpretant will involve all three types. The recursion runs only in one direction: animals that form the simplest of signs cannot "see forward" to more complex ones, but animals that form complex signs do so only by building on the capacities to form simpler ones. The hierarchic infrastructure of complex signs has been emphasized by Deacon in several accounts of human symbolism (Deacon 1997, 2012a, 2012c), and it carries large consequences not only for hominins but for all animals whose semiosis surpasses a rudimentary level.

This sort of hierarchized achievement of complex representations in the world has been viewed from other, nonsemiotic perspectives. Evolutionary linguists Sergio Balari and Guillermo Lorenzo (2013), for example, model the problem of animal representations through computational science. They posit an all-purpose "computational system" that underlies the mental activities of a wide range of animal species possessed of highly developed nervous systems. The computational system comprises two fundamental traits: a pattern-generating capacity, or "sequencer," located in the basal ganglia deep in subcortical areas of the brain, and memory, located in the cortex. These two loci, connected in feedback loops running from cortex to subcortex and back, give rise to the perceived relations that enable representations, intentionality, and the experience of aboutness. The differing levels of complexity of computation in the animal world, Balari and Lorenzo hypothesize, are chiefly a matter of differing hierarchical arrangements of cortical capacities, especially as enabled by increasing capacities for working memory. The arrangements can be roughly mapped onto Peirce's emotional, energetic, and logical interpretants, and they will help us when we turn in chapter 5 to model the emergence of cultural systems among hominins.

The range of semiosis

Understanding the thirdness of signs and the role of the interpretant suggests that we should extend semiotic capacities very far in the animal kingdom: all the way to the limit of the phenomenon of attention. It is not obvious how far this is or where to draw a phylogenetic boundary between purely informational organisms and informational-semiotic ones. Wherever we place it, it will in all likelihood be a blurry one, because the capacities that enable semiosis are multiple and not easy to discern, because they emerged gradually as the product of a long, branching evolution, and because they probably formed independently in different lineages. In spite of this uncertainty, the nature and role of the interpretant make two things clear. On the one hand, the range of semiosis will extend far beyond the human realm to include a large number of animal species. The attentional activation of the interpretant renders large reaches of the biosphere a *semiosphere* (Lotman 2005): a vast network of semiotic acts on the part of a certain large and varied group of organisms. On the other hand, this attentional aspect poses limits to the semiosphere. There is an important difference between a paramecium or perhaps even a flatworm registering a chemical gradient in a way that sets off programs that make it move (complex, evolved information processing, but still akin to that of a bacterium) and a bird that recognizes certain berries as food or another bird as a predator (sign making). Unless we think that the paramecium or flatworm can attend to food like the bird does, we cannot see either one as an interpretant-making, semiotic organism; still less can we see bacteria as sign makers, and still less than that, molecules.

Biosemioticians who take Peirce's semiotics as a starting point often miss the import of his interpretant. Without it, the sign is reduced from thirdness to secondness—an impossible phenomenon in Peirce's theory, but one implied in many biosemiotic discussions—and made indistinguishable from information. All informational processes then come to be signs, and there can be no limiting of signs to the biosphere, for if we identify signs in the conformation of DNA and RNA, it is hard not to find them also between a black hole and its galaxy. The situation is related to the predicament Deacon finds himself in by extending a capacity of "interpretation" all the way to molecules. In both cases a very important difference concerning a large group of living things has been obscured. We can call it computation, as defined by Balari and Lorenzo, or attention, or the capacity to strike up a relation to a relation, and something close to it has recently been assimilated, with a nod to Peirce, to the much-developed idea of mediation (Grusin 2015)—but in every case what is named is the foundation of semiosis in thirdness.

An aside: Why signs?

What advantage is gained by modeling organismal reactions to the world in terms of the two categories of information and signs? The semiotic approach introduces content not as an inevitable aspect of information but by virtue of thirdness, thus distinguishing signs from contentless Shannon information and reliable covariance. Because of this distinction, the differentiation of signs from information enables us to discern discrete processes and levels of complexity involved in organismal response and cognition. This is clear already in the hierarchized types of interpretant described above, and it will become clearer below, when we discuss the kinds of signs arising from interpretant action.

Still, the fundamental distinction advanced here between the information processing of all living things and the sign making of some (relatively few) animals is a blunt one, and it needs further gradation to map its categories more precisely onto the wide variety of animal responses we see in the world. In its bluntness it recalls a division that Fodor (1986) advanced in the 1980s between organisms with "mental representations" and those without them. Fodor, however, differed from the view described here on the question of content. He saw computation of some kind behind all behavior, even that of his exemplary paramecium, and he assigned content to all this computation. Such an extension leads us quickly back to the mistake of biosemioticians who conclude that even molecular covariances must be *about* something. Philosopher Ruth Millikan (1989), in a well-known response to Fodor, saw the need for distinctions among his representations, and in doing so she steered a careful course around the fallacy of ubiquitous content. If we follow it, we can see that her course, without mentioning Peirce or semiosis, converges on the idea of the interpretant.

Millikan extends mental representations farther than Fodor and does so by articulating six fundamental ways in which those of a bacterium differ from ours—in effect, a graded spectrum of complexity of the representations. She assigns some manner of "biosemantic" content to all these kinds of representations, and so her formulation does not make use of the information/sign distinction I have described. But her most important move is to shift the ground under the representations by focusing on their use, not their production. "It is the devices that *use* representations," she writes, "which determine these to be representations and, at the same time (contra Fodor), determine their content" (1989, 283–84). Millikan's content is made, in other words, in being exploited *as* content, just as signs are created in the process of making them signs: re-presentations of something else. We see a Peircean thirdness, indeed an interpretant-like action, coming into view here. "If it is actually one of a system's functions to produce representations . . . ," Millikan continues, "these representations must function as representations for the system itself. Let us view the system, then, as divided into two parts or two aspects, one of which produces representations for the other to consume." The producing aspect will make representations only in accordance with the evolved properties that enable the consuming aspect to make them representations *of something*: "we can construct a semantics for the consumer's language. . . . The sign producer's function will be to produce signs that are true *as the consumer reads the language*" (285–86). Millikan's producing and consuming aspects are not divided in reality, of course, but only for heuristic purposes. What she has described, from a vantage different from Peirce's, is the two-sided, receptive-reactive operation of the interpretant in its generation of signs.

The advantages of the semiotic approach, then, are that it marks a division in the living world between content and its absence and delineates, on the content side, a spectrum of different ways of producing and perceiving it. By localizing aboutness, meaning, and content, it addresses the obvious fact that the interactions of some animals with their environments differ in quality from those of all other organisms, however complex these may be. Semiosis, as we will see in discerning types of signs, is not an imaginary construct, and what it describes is not a set of fancies we imagine animals to indulge in. Instead, it specifies the kinds of mediated relations between things that certain organisms, endowed with neural systems of sufficient complexity, can conjure into being and react to. The constraints that govern the kinds of relations, as Deacon has taught us, are not merely those internal to the organism doing the sign making, delimited by its capacities; there are also, prior to these, constraints in the relations of things in the world that are happened upon or tapped into by animals with certain capacities (Deacon 2012a). This naming of a prioris is no small advantage in our attempt to understand animal response and learning. It can help us avoid the tautologies often evident in discussions of such matters, circularities in which the very vocabulary to be illuminated by the analysis comes to be the lexical toolkit employed in the analysis. (This vocabulary ranges widely, including notions such as consciousness, mental representations, self-awareness, subjective experience, and the "something it is like to be" a particular animal, in Thomas Nagel's famous phrase [1974].)

Three levels of niche construction

In its broad outlines, the semiotic approach brings additional distinctions to the question of niche construction. Since semiosis comes about through the interactions of perceiving organisms with their environments, it is always niche constructive. In other words, for sign-making organisms the interpretant defines some part of their ecological interactions. But it does not define them all, since these organisms overlay their semiosis on the informational base they share with all organisms. So, while semiosis is always niche constructive, niche construction is not always semiotic; indeed, across the whole biosphere it is mostly informational, without signs. If we were to imagine two concentric circles, there would be a very large one for the biosphere—all of it niche constructive and informational—and a much smaller one within it representing *semiotic niche construction.*

Where will the niche construction of cultural animals be found in this diagram? We have seen already that semiosis names the elaboration of information, in certain precincts of the animal world, beyond covariance to a state of aboutness, replete with content and meaning. This aboutness is the starting point for all culture. Knowledge taught, imitated, learned, and passed on to successive generations is always knowledge *about*, just as a sign is always a sign *of*. This is as true of cultural knowledge that appears in behavioral form—a technique for flint knapping passed along by hominins possessing no modern language—as it is for the most abstract ideational construct. A small portion of semiosis in the world results in culture, but all culture is a semiotic condition, not merely an informational one. So, in our imagined diagram, there will be a third circle within the small, semiotic one: the far smaller circle of *cultural niche construction.*

This means that we have arrived at three levels of niche construction, not just the two discerned by the niche constructionists. All of it is, as they saw, a matter of information transmitted, and some of it is also a matter of cultural transmission. But in between there exists another, middle ground that they did not notice, a noncultural niche construction that involves both semiosis and information rather than information only. This three-part model captures the difference between algae blooming in a pond (informational alone) and beavers making dams (semiotic as well as informational). At the same time it captures the difference between the beaver and a hominin teaching its child a pattern of flint knapping: an animal making its niche through an amalgam, of unprecedented richness, of information, semiosis, and semiotically based cultural transmission.

:3:

Of Peirce's several typologies of signs, the most important is also the most famous: his tripartite division of icons, indexes, and symbols. These are further keywords commanding our attention. In Peirce's mature thinking the three terms did not name all aspects of signs but referred specifically to three different kinds of relation between signs and their objects, and Peirce invented other terms to describe both the relations of signs to their interpretants and also the aspects of signs themselves that enable them to signify (Atkin 2013). I will leave aside the terminological tangle, while maintaining the important principle Peirce asserted with it: to understand signification fully, given its triadic structure, it is not enough to characterize only the sign's relation to its object. This is all the more important when we try to understand the range of sign types beyond humans, where the dual, attentional/intentional experience of an interpretant involved in each of the three main types is more limited and more sharply defined than in humans. In an un-Peircean way, then, I will employ "icon," "index," and "symbol" as umbrella terms capturing both the sign-object and the sign-interpretant relations.

Icon An *icon* is a sign related to its object because its perceiver recognizes a qualitative resemblance between the two; because of this relation, Peirce thinks of an icon as "a sign which refers to the Object it denotes merely by virtue of characters of its own which it possesses" (quoted from Short 2007, 215). This seems simple enough, and we can quickly summon human examples such as a leaping deer on a deer-crossing sign or mammoths and aurochs painted on cave walls; but in the larger transspecies perspective the icon is no easy matter to define. This is because it is the most basic and widespread kind of sign, arising from the simplest forms of organismal memory and learning—the rudiments of animal perception and attention—and hence minimally separated from sheer information processing. On one side of the divide, as we have seen, all living things are genetically programmed to register, without perception or the attention it involves, informational correspondences; on the other side, a smaller but still immense array of species construct a nascent semiosis in their recognition, through likeness, of food, predator, or conspecific (mate or rival).

Perceiving, attentive animals parse the world in many ways according to likeness or unlikeness. Thus, to give an example developed by Deacon, a bird might find a moth to eat because of the unlikeness of its coloring to the bark of the tree on which it has landed, while missing another one colored so as to camouflage it. Considering this example, Deacon judges icons to be formed in a negative way, through "the act of not making a distinction" (Deacon 1997, 74)—in this case, on the part of the bird left hungry in the presence of the camouflaged moth. But this is to limit too narrowly the interpretant making involved in iconicity. Think of a bird choosing a berry to eat. It forms an icon in a positive way, as it recognizes through learned or remembered likeness *what it eats*. In the same way it will also iconically identify a mate, track a rival, and avoid a predator. In each case, a threshold has been crossed separating the bird's activity from that of a flatworm processing chemical information to register a gradient and find its foodstuff. The crossing is marked by the presence of an interpretant and the making of an icon.

The difference between semiotic bird and informational flatworm suggests analogous differences that are less stark and more difficult to judge. To speak of a single phylum, mollusks, it is clear that the cephalopods (octopuses, squids, and cuttlefish) are attentive creatures that interact with and learn from their environments in very complex ways (Godfrey-Smith 2016). They are semiotic creatures, capable of relating things outside themselves to one another and of iconism at least. It seems equally clear that clams and other bivalves, also mollusks, do not have these capacities but are informational creatures only; while the case of gastropods (snails and slugs) stands somewhere between these (Perry, Barron, and Cheng 2013). The instance of cephalopods, taxonomically far from vertebrates, is one of several that can be adduced to show that a semiotic relation to the world emerged independently in separate phyla. This convergence might argue in favor of its being a natural outgrowth of neural systems of a certain level of complexity faced with the a priori relational constraints mentioned above.

Index An *index*, the second of Peirce's three main sign types, arises from perception of a causal relation, a pointing indication, or a contiguity between sign and object. The wide breadth of indexicality is signaled here, and it is also repeatedly implicit in Peirce's definitions. In 1901, for example, he described the index as "a sign . . . which refers to its object not so much because of any similarity . . . as because it is in dynamical (including spatial) connection, both with the individual object . . . and with the senses or the memory of the person for whom it serves as a sign" (quoted from Short 2007, 219). Almost any kind of perceived connection of sign and object other than resemblance can form an index. The famous example is smoke, considered as an index of fire.

Indexicality is a more complex kind of signification than iconicity, setting a higher bar of cognitive prerequisites, and is less widespread among animals. This is because the relation of sign and object in it has been distanced and in a way doubly mediated. An animal that can interpret a rustling branch as a predator achieves a semiosis more advanced than one that can only recognize a predator in the flesh. The interpreted branch is an index of an object that might not be present and, in any case, is not visible at the moment of interpretant making, so the forming of the sign must rely in a complicated way on prior experience. The sign-object connection at the present moment springs from a learning experience or memory of other similar circumstances in which branch and threat were associated. In a similar way any pointing gesture requires a learned history of other pointing gestures to be interpreted. The index, then, arises from a matching of the earlier situation (which juxtaposed rustling with predator) with the present one (perhaps incomplete: no predator after all, just the breeze). This matching reveals an iconic element in the index and so discloses the basic hierarchic connection between the two sign types: the interpretant relies, in effect, on an iconic resemblance of likeness between the two situations, the remembered one and the present one, in order to form a new sign, the index (Deacon 1997, 2012a). Indexicality will in this way always rest on a foundation of iconicity. In its double mediation—sign-object bond in the present mediated by the interpretant, and present sign mediated by another learned or remembered sign-object pair—it requires more advanced modes of perception and cognition, among them a situational memory and a significant capacity to learn from experience. These suggest that indexicality relies on an advance in the complexity of the neural system involved or, in terms of the computational model Balari and Lorenzo (2013) have described, in the kind of computation available.

Symbol

A *symbol*, finally, is related to its object through the operation of governing laws or conventions, which the interpreter requires as a background in order to connect the sign to its object. As Peirce put it in 1903, a symbol is "a sign which refers to the Object that it denotes by virtue of a law, usually an association of general ideas, which operates to cause the Symbol to be interpreted as referring to that Object" (quoted from Short 2007, 220). Customary examples of symbols come from human language, where the individual word is the symbol, and grammar, phonetics, morphology, and other general aspects form the laws or "association of general ideas." This example in itself suggests that we might expect to find a very narrow dispersion of symbolism in the animal kingdom, and indeed it is arguably limited in the biosphere today to humans and a few individual animals they have trained.

It is important to understand the operation of the laws, conventions, or associated ideas on which symbols depend. These, as Peirce's definition specifies, do not operate on the object, and they do not extend between the sign and the object, somehow constructing that connection. Instead, they structure the interpretant, organizing a network of symbols and creating systematic relations among them—combinatorial grammars, for example—that are independent of the objects in the semiosis. The individual symbol takes on its capacity to be linked to an object *outside* the system through the perceiver's understanding of its relations to other symbols *inside* the system. Individual signs within the system function not merely as independent symbols but also as pointers to symbols around them—indexes, in other words, specifying the reference of those other signs. And through this mutual, index-symbol relation the system as a whole comes to function as an index pointing to arrays of objects outside it—the world of objects and actions outside language, for example (Deacon 1997, 2012a). We can see why it is hard to imagine that this kind of semiosis extends much beyond *Homo sapiens* today. How far it reached in our extinct hominin relatives is a fascinating question, but attempts to answer it have too often started from an inadequate understanding of symbolic semiosis itself (Tomlinson 2015; see chapter 7 below).

Symbols require systems

A foundation of indexicality is thus essential to the symbol, and this extends the hierarchy of sign types Deacon has pointed out: just as an index depends on iconism, so a symbol depends on indexicality. The systematization of the array of symbols is also fundamental to symbolism. It is what enables individual symbols to function as indexes, pointing to the signification of other symbols around them and creating a mega-index from the system as a whole. Symbolism requires the enlargement and complication of the mediated distance of sign from object that is opened in the index in more modest form (the predator that might not be present behind the rustling bush). The widened distance comes from an interpreter endowed with faculties that enable the systematic arrangement of sets of signs. Systematicity is the heart of the symbolic interpretant, and it probably is founded in additional levels of complexity in neural system and computational cognition (see chapter 5).

It is hard to overstate the importance of these issues in thinking about the rise of human modernity. The gradual systematizing of hominin culture is something we can track through many proxies in the archaeological record, so we can follow the step-by-step emergence of semiotic complexity, watching its growth in organization and gauging the increasing intricacy of mediation between sign and object. This enables us to propose approximately when in our biocultural evolution symbolism coalesced from highly developed modes of indexicality. Moreover, we can also discern in-between semiotic practices where indexes themselves take on some degree of systematic organization, even though they do not wholly depend on it for their semiosis. This kind of *hyperindexicality* (as I will call it) can be highly developed, and it remains basic to human culture today, where instances of it include, in certain of their fundamental aspects, music and ritual (Tomlinson 2015; Silverstein 2003). Such possibilities suggest an unbroken gradient from indexicality to symbolism, and I will argue in chapter 7 that symbolism arose, in our deep history, by smooth gradations from earlier indexicality and requires no dramatic leaps in the cognitive capacities of ancient humans to explain it. When it appeared, symbolism carried explosive potential, but it needed no explosion for its appearance.

: 4 :

Information, semiosis, culture When Maynard Smith and Szathmáry (1995) proposed their list of the major transitions in the history of life (see p. 1), they also described features that tend to recur in them. One of these is a shift in the nature of information transmission, as, for example, when genetic materials came to be recombined in sexual reproduction. We can now specify two additional shifts in information transmission: semiosis and culture. These entail rearrangements of the pathways of information, and they also involve innovation in the kind of thing that moves along the pathways. In evolutionary history, semiosis required innovations in the capacities of living organisms. The emergence of signs from information was a momentous change, enriching a biosphere of correspondence with aboutness, content, and representation. The difficulty of specifying where in phylogenetic history this threshold occurred should not lead us to underestimate its importance. We can see, in the light of the processes described above, organisms on each side of the divide: those living by information processing alone and those that additionally make interpretants.

The second new dimension we can now add to Maynard Smith and Szathmáry's information is culture. What are the general differences between the semiosis that is widespread in the animal world and the much rarer elaboration of semiosis that constitutes culture? What are the features that have enabled a few animal taxa to elaborate semiosis into culture? Such questions can easily exhaust themselves in debates about the extent of animal culture in the world today. These are of immense inherent interest, of course, and they have greatly raised our awareness of the complexities of nonhuman animal behaviors. They suggest that we should draw the borders of nonhuman culture liberally, to include at least a small range of mammalian and avian lineages: certain primates, some cetaceans, a few other mammals, and some birds. All the same, we must be careful not to confuse animal culture with the far broader category of animal sociality. Ants have complex societies, but they do not have cultures. Many instances of highly developed avian and mammalian sociality also exist without giving rise to culture. Baboons offer a well-documented example, with intricacies of social relations in their troops that can easily be mistaken for culture.

Social versus cultural semiosis

What, then, differentiates cultural sign making from sign making in general? Imagine a few contrasting examples of animal behavior. In the social interactions of a baboon troop, signs are everywhere: welcoming signs, warning signs, aggressive and submissive signs—all of them bodily and vocalized gestures used to negotiate a hierarchy of matrilinear power, to protect babies from adult males, to rank-order access to food, to protect the troop from predators, and so forth (Cheney and Seyfarth 2007). Most of these signs are indexes, some of them are icons, and none of them are symbols; but of whatever type, signs are the medium of baboon sociality. The social relations that baboons negotiate through their signs unfold as face-to-face, here-and-now situations, and the signs apply to the momentary flux in which they are deployed. There is, at most, only a minimal displacement by which signs extend their temporal reach away from the transaction at hand. This is a kind of sociality that anthropologist Maurice Bloch (2013) has called the *transactional social*, essentially limited to a situation of copresence and a real-time temporality.

Here I must be clear and careful. In maintaining that there is little or no displacement of signs away from their situational use, I am not maintaining that the baboons do not *learn* from the situations and the signs in them—learn, for example, that one adult female is more powerful than another—and then *remember* their lessons in future interactions. Baboons are smart animals, with well-developed episodic (or situational) memory and abundant learning of several sorts: associative or conditioned, contextual, operant or instrumental, and perhaps more (for these types, see Perry, Barron, and Cheng 2013). In social and semiotic terms, their interactions rise to a high level of nuance and indexical complexity. Even so, all this does not amount to culture.

Now think of some other, contrasting situations, all of which have been put forward as examples of nonhuman culture. A young songbird learns from adults around it a set of elements from which complex songs can be constructed, elements specific to its species and also to a local population (a "dialect"); the bird modifies these elements slightly and passes on an altered dialect to succeeding generations (Freeberg, King, and West 2001; Williams et al. 2013). A pod of humpback whales communicates through highly structured calls specific to the group, which are altered and further developed each year when the pod gathers in warm waters to mate and calve (Payne, Tyack, and Payne 1983; Eriksen et al. 2005). A monkey devises a new behavior, washing sand off a sweet potato before eating it; younger monkeys in the troop repeat the action, and it is passed on through generations (Kawai 1965; Hirata, Watanabe, and Kawai 2001).

In all three of these situations a new ingredient has entered into the use of signs. They have taken on a temporal displacement from the face-to-face interaction in which they were first deployed. This displacement allows something more than what is evident in the baboon example, more than a learning of the dispositions or powers or social ranks of other baboons. There a lesson was given and remembered. The object of the sign was learned, but not the sign itself. It was an indicator of the needed lesson, bound to the transaction of which it was part, registered as a conveyance of content, and discarded in the ongoing flux of social interaction. In the cultural situation, on the other hand, *the signs themselves are learned*, whether songs, calls, or even material procedures. A new abstraction appears by which signs are distanced from any particular situation, readied for repeated application in future situations, released from their here-and-now proximity. The abstraction recalls the old adage "Give a man a fish, and you feed him for a day; teach a man to fish, and you feed him for a lifetime." Most social animals, in their semiotic acts, are handing over fish; cultural animals are teaching how to fish.

This semiotic abstraction marks the appearance of culture. We have already defined culture broadly and loosely as knowledge acquired through social interaction with conspecifics and passed along to future generations. Examination of semiosis allows us to refine our sense of the second element of this definition, the transmission to others of what has been learned. This requires the abstraction of social learning from its initial context by means of semiotic displacement. In Bloch's terminology (2013), the limitations of the transactional social have been transcended, bringing about what he calls the *transcendental social.* Culture is the product of a transcendental semiotic condition that cuts signs loose from the circumstances of their use.

The transcendental interpretant

The displacement introduced into cultural semiosis is not a distance opened between the sign and its object—the kind of distance or mediation we have already described in the sign types of index and symbol, which arises in each case from their hierarchical reliance on less complex signs. Instead, this new distance in the semiotic process intervenes between the sign-object-interpretant relation as a whole and a particular use of the sign. This opens a new horizon for social learning, the end point of which is now doubled, involving both a focus on the object of the sign in its immediate context and the carrying forward of the sign to other, new contexts. This opening requires additional complexities in cognitive computation, and with them a new function for the interpretant appears, a transcendental operation that releases the sign from an attachment to one situation. Culture stems from this redirection of the interpretant and from the novel abstraction it introduces.

In the deep history of hominins we can follow the increasing power of this transcendental interpretant in a slowly advancing "release from proximity" that archaeologists have tracked across the last 500,000 years (Gamble 1999; Rodseth et al. 1991). Early examples include the transport of materials for future use rather than discarding them in situ, the extended curation of tools across multiple uses, and (probably a somewhat later development) the exchange of materials between groups. These behaviors and others like them outline the growth of abstraction in our lineage—a history of the coalescence of the transcendental social, in other words. This, I will argue, is connected to the systematization of culture over the same period, which led finally to symbolism: it too was a mode of cognitive abstraction. In chapter 5 we will see how this release was connected to the general features of the highly developed culture of late hominins: traditions, archives, systems, and epicycles.

: 5 :

Foundations of hominin culture From informational covariance to transcendental culture—we have moved far in this chapter, and it is well to recall the proposal with which I began: that cultural niche construction was not merely important for late hominins but finally so powerful a force that it changed *in kind* the feedback patterns of genes, environment, and culture in human evolution. The full exploration of this proposal in later chapters will be built on a foundation of the feedback and feedforward dynamics outlined in chapters 2 and 3 and the semiotic concepts introduced above.

At the bottom of this foundation is the link, in niche construction, of cultural and genetic inheritances through the mediation of the environment. To describe how this works—that is, to add qualitative terms to the mathematical ones surveyed in chapter 3—requires understanding the semiotic emergence of animal culture itself. This starts from the animal attention that creates signs, meaning, and representation in the world, by virtue of the semiotic master key, the receptive-reactive interpretant. Culture relies also on animal capacities to create signs of different types, capacities whose rarity increases with the growing complexity of the type in question. The icon is pervasive wherever animal attention exists. In the presence of additional cognitive capacities it gives rise to the index, more limited in its dispersion but still widespread. The index in turn gives rise to the symbol through another, far rarer coalescence of cognitive faculties, probably limited to the hominin lineage. Symbolism and also some hyperdeveloped indexicality rely on an interpretant maker able to command systems of signs, and these systems played a lead role in the burgeoning cultural horizons of ancient humans (see chapter 5).

Culture depends also on one more elaboration of these semiotic operations: the abstraction of signs from individual situations of their use. This transcendentalizing of semiosis enabled the passage of acquired knowledge from one generation to the next. It extends beyond humans to appear wherever rudimentary cultures take shape: among the songbirds, whales, and monkeys adduced as examples above, and in a few more animals. All these nonhuman cultures, however, subsist on icon and index alone. Only in humans does the sign-object abstraction join forces with the powers of hyperindex and symbol.

The mortar that holds together this foundation of late hominin culture is animal *learning*. It forms an implicit part of our basic definition of culture, a prerequisite for the acquisition of knowledge during a lifetime that is passed along to others. Most types of learning do not in themselves give rise to cultural processes but are instead more general aspects of the interactions of many kinds of animals with the world. Habituation or sensitization to repeated stimuli, stimulus-response conditioning, operant or instrumental learning about the results of actions in the world, and even learning from episodic events in specific situations are all widely dispersed, and they were securely in place in hominins from the beginning of the lineage. So were the capacities these kinds of learning in their turn depend on, especially short-term procedural memory and some situational or episodic memory. These things we can take for granted in all hominins.

Social learning in culture

Social learning—learning from conspecifics in face-to-face interaction—raises more complicated questions. It too is widespread, though not so much as other, more basic types. It does not always give rise to culture, and the example of baboons shows that it can exist in highly developed, intricate, but noncultural forms. Social learning linked to cultural semiosis—a fully cultural social learning—is less widespread still, but it reaches far back in the hominin lineage; certainly the hominins producing the first Acheulean hand axes 1.75 million years ago showed it in developed form. Confronted with hominins much later in the Lower Paleolithic period—for example, a band of *Homo heidelbergensis* in Europe half a million years ago—we are unmistakably in the presence of complexly cultural creatures; for this the archaeological evidence from a scene such as the communal butchering at Boxgrove, England, is conclusive (Gamble 1999; see chapter 7 below). By the time of the advent of Neandertals, some 350,000 years ago, or of our own species of "sapient" humans, perhaps as little as 50,000 years later, cultural social learning was an old story.

This deep heritage of cultural social learning does not mean, however, that there were not momentous changes along the way. Archaeological findings point to a stunning growth in the richness and complexity of social learning among hominins across the last half million years and even across the last 200,000 years. To understand how this shift came about, we need to outline the general prerequisites for social learning, starting again from the phenomenon of attention.

For the most part, animal attention in the world is a solitary phenomenon. A stimulus rises to salience in the animal's perception and it reacts. Two animals or more might attend to the same stimulus, but they do so in a solitary fashion, jointly but singly. Cultural animals, even in rudimentary cultures, do something more than attending jointly: they share attention with one another. What is the difference between joint and shared attention? According to primatologist Michael Tomasello, joint attention of a highly elaborate sort can be witnessed in the phenomenon of chimpanzee group hunting. When the group tracks a monkey, each member attends to the same thing, the prey, but each does so only to further its own access to it, without any cooperative motive or strategy. Shared attention instead involves two or more animals focused on the same thing and with a common, superindividual end in mind; it thus rises to a condition of shared intention as well as attention (Tomasello et al. 2005; Tomasello 2008). A scene of an adult *Homo ergaster* one million years ago teaching a stone-knapping technique to a younger member of the group must already have presented a categorical advance on chimpanzee hunting. The hominins involved were working cooperatively toward the same end and were not constrained within the limiting pursuit of individual advantage.

Theory of mind Tomasello calls the cognitive capacity undergirding this cooperation "recursive mind reading," but it is more often known as *theory of mind* (Carruthers and Smith 1996; Bloch 2013). This phrase refers to the ability of an animal to recognize in another animal, usually a conspecific, awareness like its own of the outer world and events. Its cognitive faculties enable it to see that the motives and intents of the other animal resemble its own motives and intents—that there is a similar mind at work. Like other capacities discussed above, theory of mind reaches beyond humans, though it is probably rare in the animal world as a whole; as with those other capacities, it is difficult to judge exactly how far it extends. It exists in differing degrees of intricacy. A monkey might recognize similar goals in another monkey, but it seems not to be able to see that the other monkey recognizes it also; it lacks second-order theory of mind. Humans carry theory of mind easily to a third order—I recognize that you recognize that I recognize you as a similar kind of creature—but turning the screw beyond this, to a fourth or fifth order, we manage only with effort. This back-and-forth mutuality in human theory of mind is what led Tomasello (2008) to call it recursive.

Theory of mind underpins cooperation because it can transform joint attention into shared intention. A joint activity of two animals becomes truly cooperative when a second-order theory of mind is involved, each seeing the other pursuing the same goal (first-order theory of mind) but also understanding that the other sees the sharing in the same way (second-order). The hunting chimpanzees, as interpreted by Tomasello, see others pursuing the same prey, but they seem not to see those others seeing them back; rivalry rather than true cooperation guides the hunt. For humans, on the other hand, second-order cooperation reaches far back in our history. The scene described above of a *Homo ergaster* passing on a toolmaking skill to its young probably depended upon it, with adult and child both perceiving that the other understood the shared aim of the activity. Archaeological study of toolmaking traditions suggests that this kind of scene was enacted innumerable times across the last million years, at least.

Observation, imitation, and practice

Cultural social learning seems to require this second-order theory of mind, while simpler social learning does not. And, once such theory of mind was in place, a cascade of other learning complexities came to be linked to it. The Paleolithic toolmaking scene above depended on observation, as the young hominin took in a new skill by watching it being performed. The observational component was probably especially important in such early instances, because *Homo ergaster* and even *Homo heidelbergensis* possessed nothing like a modern, propositional language with which to explain actions and intents. Second-order theory of mind filled in for explanation, creating a gestural pedagogy as the learner observed and recognized the informing intent of the teacher's actions. It is a frequent error in thinking about ancient hominins to suppose that all complex cultural traditions required propositional language.

The learner in this scene required also the abilities to imitate and practice. Evolutionary psychologist Merlin Donald (1999, 1991) has analyzed these abilities and laid out the minimal requirements for each. They entail, first, the basic capacity to initiate voluntarily a sequence of motor actions. This was certainly already highly developed in *ergaster*, and it relied on a set of more ancient capacities: the procedural and episodic memory already mentioned and more. In cooperative situations the actions are triggered by theory of mind itself, the understanding of the shared intention pointing toward a goal, but they need something more for learning to take place: a feedback loop of rehearsal and assessment. The steps in this loop move, after the initial observation, to self-performance of the actions observed, then to observation of the results of this performance and remembrance of them, and finally to renewed performance with one or more aspects modified as necessary to achieve the desired end. Observational learning, this schematic makes clear, is directed both outward, toward the teacher imitated, and inward, toward the learner.

The element of practice in the rehearsal loop, Donald argues, is a crucial distinguishing feature of hominin learning, not found in other primates in the world today; baboons have been known to push rocks down hillsides to fight with other baboons, but, as Donald notes, no one has ever seen a baboon target-practicing. Many ethologists might extend the rehearsal loop to some other animal situations—birdsongs and whale calls come to mind as likely examples—but it is in any case an exceedingly rare conjuncture of capacities. Meanwhile, the rehearsal loop needs to be distinguished from play, a widespread phenomenon among mammals and some other animals. Play might be regarded as a kind of practice, but it seems not to be subject to the voluntary control involved in the hominin rehearsal loop. Lion cubs fighting with each other replicate and perfect instinctual patterns of behavior, but their play affords little or no opportunity for reshaping socially transmitted interactions.

: 6 :

The verge of human modernity A quarter million years ago, in a Neandertal encampment in Europe or an African gathering of pre- or early sapient humans at a natural stone shelter, niche-constructive lifeways had assumed new dimensions and unprecedented variety. These hominins commanded a semiotic intricacy outstripping anything in the nonhominins around them. Their communication showed a range of signs broader than the most complex nonhuman communication in the world today, deploying with both body and voice an array of indexical gestures. They reshaped their material environments in novel ways, resulting in sophisticated technologies, of which the lithic remains today are doubtless only a very partial sample; and they passed these technologies along to their offspring in robust traditions, employing intentful social pedagogies. The archaeological evidence suggests also that their patterns of movement across the landscape and their harvesting of its resources were shrewd, involving some measure of foresight and planning. Their social organization probably reflected a similar high development in its own complexities.

And yet: these were humans without any well-developed symbolism and with little of the hyperindexicality I exemplified above with music and ritual. They possessed no modern human language, and their technological traditions, for all their richness and skill, changed slowly through the generations, seeming from our vantage to shift little over many millennia. Their marshaling of resources, for all the foresight it had begun to show, was still mainly opportunistic. And their release from the proximity of face-to-face interactions in the here and now was far from the later attainments that would enable storytelling and mythmaking and bring into view invisible realms, hidden forces, and counterintuitive deities.

These statements are of course not established facts but inferences compelled by the archaeological evidence we have. They pose for us the fundamental problem these first chapters have prepared us to tackle: the question of how the nature and achievements of these humans were radically transformed into the modernity that was already taking shape a scant 150,000 years after our hypothetical scenes, and that would advance quickly in the millennia after that. The answer to this question involves niche construction and its feedback loops enchaining culture, genes, and environment. It involves the formation of enriched semiotic processes of greater variety, flexibility, and abstraction than before. And, most important, it involves cultural systems and the emergent powers these unleash.

CHAPTER 5

SYSTEMATIC AND EMERGENT CULTURE

:1:

Toolmaking Not many years ago, the use of tools was thought to be a distinguishing feature of humans, almost as unique to our lineage as language. Now we know better. Tool use is widespread among nonhuman animals (Shumaker, Walkup, and Beck 2011). Chimpanzees employ sticks to probe tree trunks and termite mounds for insects, capuchin monkeys put a nut on a stone "anvil" and break it open with another stone, orangutans hold leaves to their lips to alter their calls, New Caledonian crows impale insects on twigs they have trimmed for the purpose (Chappell and Kacelnik 2002; Bluff et al. 2007), bottlenose dolphins wear sponges on their beaks to protect them as they forage for bottom-dwelling fish (Smolker et al. 1997)—and the list could be much expanded, ranging even beyond mammals and birds (for a tool-wielding octopus, see Finn, Tregenza, and Norman 2009). Tool-using animals show a range of practices in obtaining and shaping their tools, and they even are known to join two or more tools together for the same task. Much of this tool use is not a species-wide trait, and in some instances it seems to be limited to a single community of the animals involved. It is not genetically programmed in any narrow sense but instead is learned, acquired behavior, and sometimes it is passed along in cultural traditions of technology.

Given this widespread use of tools in the nonhuman world today, it has ceased to be a surprising fact that toolmaking is a very ancient feature of the hominin lineage. Stone choppers of the sort the Leakeys unearthed in the Olduvai Gorge are now thought to reach back 2.7 million years (Leakey 1971; Klein 2009). They were produced either by the earliest hominins or by now-extinct lineages of australopithecines—or by both. Oldowan stone tools are presumed to be the fragmentary record of technologies involving

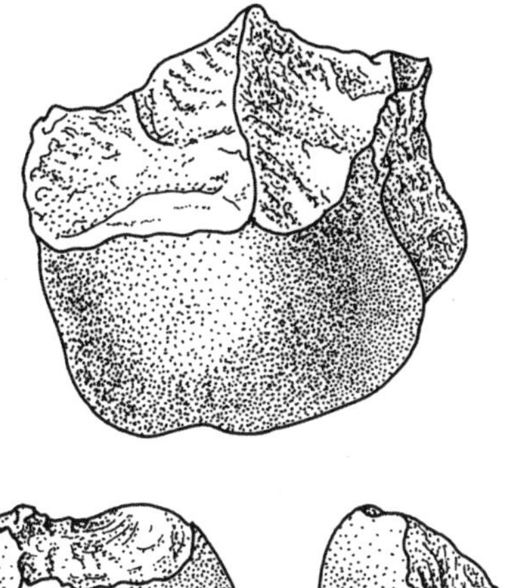

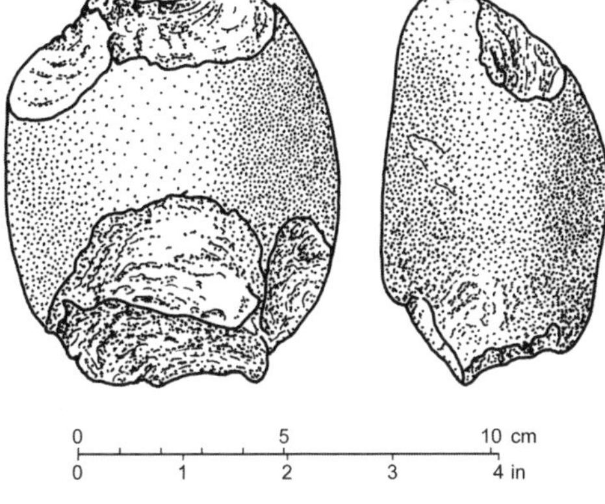

Figure 1. Oldowan choppers, redrawn by Virge Kaske after M. D. Leakey, from *Olduvai Gorge: Excavations in Beds I and II, 1960–63* (Cambridge: Cambridge University Press, 1971).

other, more perishable, minimally processed materials that have not survived: grasses, reeds, sticks, and the like. It seems probable that tools made from such easily manipulable materials might reach back much farther than crafted stone implements, perhaps all the way to the common ancestors of humans and other present-day apes, five to eight million years ago.

In their manufacture, Oldowan tools are not simple, and they come close to outstripping the limits of complexity in nonhuman toolmaking today—if they do not positively do so (Klein 2009; Wynn 2002). They assume two basic forms: stone cores with sharp edges and the flakes broken off the cores in fashioning the edges. Both might have been used as tools. To knap cobbles in this way, their makers had to choose a core and a stone hammer, gauge the materials in hand, and join together a short sequence of calibrated blows to the core. Often two or more of these blows were struck adjacent to one another, resulting in characteristic overlapping scars left on the core where flakes were removed (see fig. 1). Apes today can be taught to chip flakes off a core, though there is no evidence that they do it in the wild; but they do not manage this adjacency even with human encouragement.

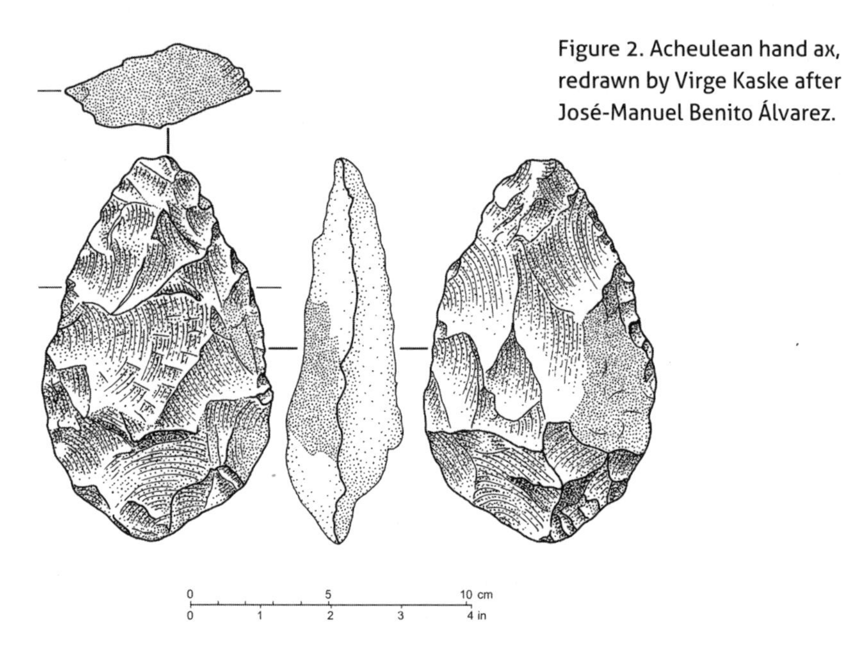

Figure 2. Acheulean hand ax, redrawn by Virge Kaske after José-Manuel Benito Álvarez.

Early toolmaking systems

While they are not simple, Oldowan tools are, however, not *systematic*. What does this word mean in the Paleolithic setting? The Oldowan tools were in all likelihood part of a cultural heritage of their makers—which is to say, they were the result of acquired behaviors passed along from generation to generation by observation and imitation of the sort described in chapter 4. The techniques transmitted were the striking of one stone on another so as to fracture it, the additional nicety of adjacent blows, and perhaps even the selection, in a general way, of core and hammer; but they do not seem to have involved any process or organized arrangement that would have resulted in a standardized design in the tools produced.

The first, rudimentary appearance of such standardization postdates the Oldowan industry and is a characteristic feature of the next major Paleolithic industry, the Acheulean, starting from about 1.7 million years ago (Klein 2009; Wynn 2002). Its most typical tools, hand axes (or bifaces), show from the earliest examples a tendency toward symmetrical design. There are mysteries concerning how this symmetry was achieved and when it arose (McNabb, Binyon, and Hazelwood 2004), but eventually it grew to be very pronounced in the industry, and Acheulean bifaces took on a fine degree of consistent, replicated design (see fig. 2). The standardization of design suggests some standardization also in the process of manufacture behind it, a regularity in the sequence of operations performed on the cobble to produce the tool. This "operational sequence" (or *chaîne opératoire*, as it is often called, acknowledging the French archaeologists who developed the concept) was what was imparted in some general way as the production of bifaces was passed from generation to generation. It exemplifies in simple form a *cultural system*: an organized array of signs and behaviors connected to them that comes to be transmitted as a whole in cultural fashion, by social, observational learning.

Now, it could be argued that even Oldowan tools reflect a rudimentary system. The systematicity of culture is another phenomenon that we must be content to exemplify across a graded spectrum, realizing that the exact placement of a border to mark system from its absence is not practicable. The situation recalls the borders between information processing and the most basic, iconic semiosis or between animals with rudimentary culture and those *almost* with it. As in those cases, it is not important for my purposes to determine exactly where to draw the line between systematic and nonsystematic culture. But it *is* important, on the other hand, to take account of the increasing complexity of the systems themselves. What appeared in the Acheulean industry was an array of operations more intricate than in Oldowan practice, and these had to be transmitted to achieve consistency of biface design. If we have not crossed a Rubicon here, we have at least moved along the spectrum.

Levallois systems

To see where this repositioning will carry us, let us look ahead to another example, a Middle Paleolithic one dating from 300,000 to 200,000 years ago, still long before human modernity. A third major lithic technique discerned by archaeologists, widespread in this period through Eurasia and Africa, probably practiced by several species of hominins, and firmly associated with Neandertals, is called Levallois. (In its association with Neandertals it is sometimes also referred to as Mousterian.) Levallois technique has been studied as intensively as anything else in Paleolithic archaeology, and it has been divided into subcategories according to distinct regional practices and differences in technique and product. But a common feature of Levallois tools is the intricacy and organization of the operational sequence employed to produce them (Dibble and Bar-Josef 1995; Boëda 1995; Chazan 1997).

Figure 3 renders the basic process in schematic form. First, the edges are struck off a cobble, roughing out a core of the desired size and shape (steps 1–2; because of these preliminary steps, Levallois is called a "prepared-core" technology). Then, using the edges created by this first flaking as multiple striking platforms, flakes are chipped off the top of the cobble to form a convex surface (steps 3–4). Next, one end of the cobble is taken as a striking platform for the production of the desired point or flake (step 5); this plat-

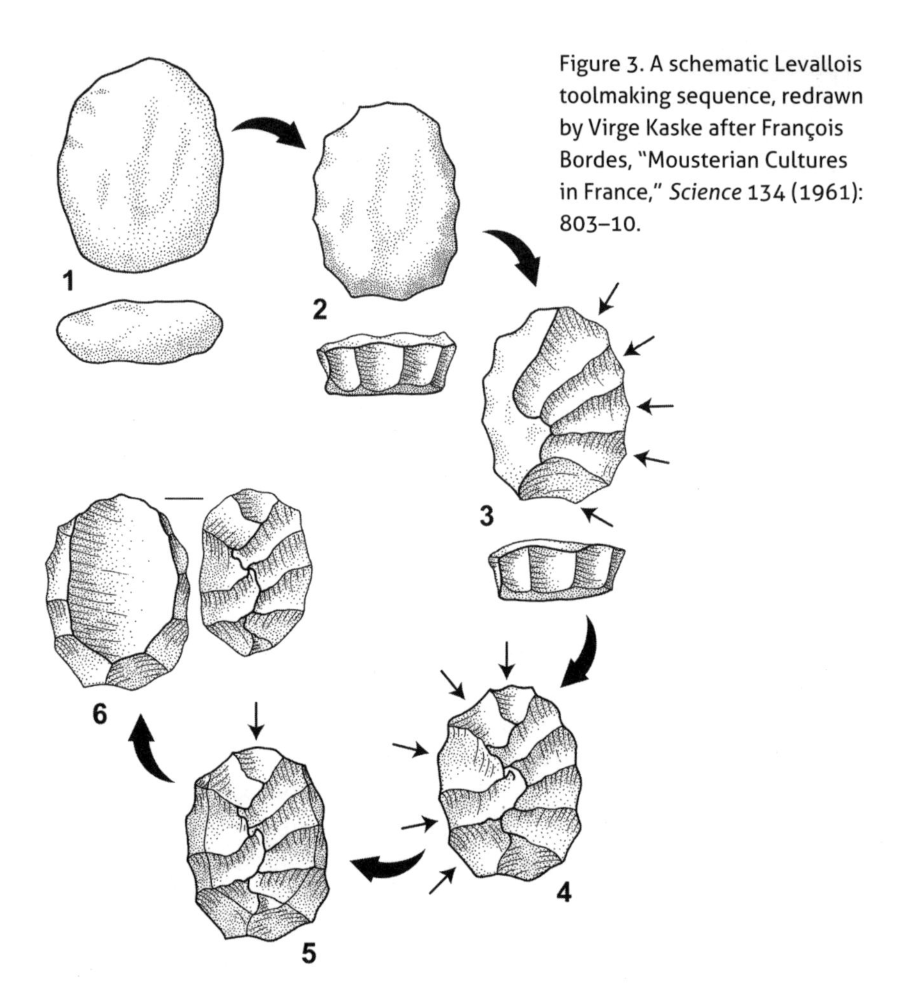

Figure 3. A schematic Levallois toolmaking sequence, redrawn by Virge Kaske after François Bordes, "Mousterian Cultures in France," *Science* 134 (1961): 803–10.

form itself may be prepared by lopping a flake off this end. A single strike at the proper place on the platform will flake off an implement (step 6) either finished or requiring minimal retouching (not shown).

This Levallois operational sequence reveals something new, above and beyond Acheulean and Oldowan procedures: an intricate internal organization to the sequence. Its individual operations take on functional distinctions from one another to become several *types* of operations, including those preparing the core, those forming the convex surface, the one striking off a point or flake, and those needed for retouching it. These types form a temporally organized hierarchy of gestures, since to achieve the desired result they must be followed in order and since the later ones cannot become practicable until the earlier steps have been successfully performed. Such functional and temporal hierarchization is not evident in Oldowan or even in Acheulean production. As we think about the transmission of Levallois technology through generations, we need to imagine the conveying of the practice thus organized *as a whole*. The social learning in this case depended on taking in the integrated set of operations, any of which, practiced piecemeal, would not lead to the desired outcome. Levallois is a full-fledged cultural system, and its functional differentiations and hierarchic organization are two hallmarks of such systems.

Composite tools

Levallois knapping is not the only technology from the Middle Paleolithic period in which we find these hallmarks. Evidence has grown conclusive in recent years that hafting techniques, by which stone blades were affixed to wooden handles, reach well back into the Middle Paleolithic among Eurasian Neandertals as well as presapient and early sapient humans in Africa (McBrearty 2007; McBrearty and Brooks 2000; Langley, Clarkson, and Ulm 2008). A hafted tool—a spear, for example—is a composite construction with several distinct parts, each produced through its own operational sequence. The stone point needs to be prepared, usually through Levallois or similar techniques; the handle must be fashioned and notched or otherwise modified to receive the blade; hide or grass straps to secure the blade must be produced or gathered; and adhesive substances needing mixing and heating are also a part of the process—the most complex production of all. All these then are joined to form the tool, which is in effect a materialized cultural system, gathering its parts, functionally distinct, into an organized whole. Here too hierarchic thinking is involved, not only in the production of the individual parts but also in the parts-to-whole design of the tool: a compositional or containment hierarchy, with component parts nested under a higher grouping. And—an obvious point, but an important one—the resulting tool can fulfill its function only through the achievement of this higher unity, doing things that none of its components could do alone. Imagine trying to kill a fleeing deer not with a spear but by throwing a stone point, a stick, or a wad of hot adhesive!

This small story concerning Oldowan, Acheulean, and Levallois technologies brings us to the point in deep history where my own account of human culture begins. By the time of the Middle Paleolithic period, hominins had come to transmit the practices and knowledge of their cultures at least partly in holistic systems, and these were defined, as all systems are, by the organized and more or less intricate hierarchic arrangements of components differentiated in function. Before this time, cultural transmission took less organized forms; its systems were less integrated and their components less fixed in functional interrelation. Looking back from a present-day vantage, systems disappear somewhere in the Lower Paleolithic period in the haze of a more generalized, less structured cultural transmission. From this point on back, our ancestors probably transmitted culture in ways not dissimilar to the cultural transmission of some nonhuman animals in the world today.

System in niche construction The growing systematization of culture among Middle Paleolithic hominins constituted a powerful force in their shaping of their lived environments, as the example of their novel technologies suggests, and so it brought a new dynamic to the phenomenon of niche construction. In effect it added to it another level, which we can picture as a fourth circle in the diagram of concentric circles developed in chapter 4. From the largest circle to the smallest, we moved from informational to semiotic to cultural niche construction; now, within the last circle, we discern one even smaller, representing *systematic* cultural niche construction. This raises a question that will be central to my account: How did the rise of cultural systems in the Paleolithic era alter the feedback patterns evident in more general types of niche construction?

It is important not to misrecognize this systematization as the beginning of symbolism among humans. Symbolism depends on systems of signs, as we have seen, but the converse is not true: systems do not depend on symbols or always lead to them. Systematic niche construction is semiotic, like all culture, but it is not always symbolic. The Middle Paleolithic examples are again instructive. There is no compelling evidence that any hominin cultures 300,000 to 200,000 years ago were habitually symbolic ones—that is, possessed of modern language or any arrays of signs systematized among themselves through conventional laws and deriving meaning from their interrelations. Whatever systems these humans created were elements in nonsymbolic cultures. Their toolmaking systems, at their most complex, are instances of what I have called hyperindexicality, a pre- or protosymbolic arrangement of indexes into systems simpler than those that would eventually underlie symbolic behavior and cognition. And we can speculate that this kind of borderline system characterized not only their toolmaking but also their protolinguistic communication and their social structures; for why would creatures capable of using such systems to their advantage restrict them within the borders of stone technologies? In the world today there are instances of modest systematization in nonhuman animal culture—for example, some whale communication and birdsong—and while these do not approach the complexities of Middle Paleolithic hominin cultures, they can be understood similarly as organized indexicality without the full portion of mediation that constitutes a symbolic system.

: 2 :

What were the new capacities that enabled late Middle Paleolithic hominins to systematize their cultures in unprecedented degree? What was the "value added" that Eurasian Neandertals and late presapient Africans—and perhaps some other groups of late hominins—introduced to a world already in some degree remade by the meagerly systematic cultures of earlier hominins? An answer to this question might start from several places: from archaeological evidence of the kind just now adduced or from the increasing degree of semiotic complexity or intricacy of sociality reflected in this evidence, and I will return to these later. But a bottom line of sorts can be approached by contemplating the novel cognitive capacities that subserved these other developments.

Cognitive computation

In chapter 4 I introduced the work of evolutionary linguists Sergio Balari and Guillermo Lorenzo (2013), who describe a "computational system" underlying the mental activities of a wide variety of animal species with well-developed central nervous systems. A closer look at their model can help us picture what was involved in the growing systematization of late hominin culture.

Balari and Lorenzo's central concern is the evolution of human language, and in particular its computational aspects—the Chomskyan processing that has loomed since the 1960s as a primary puzzle to be solved in hypotheses about the rise of language. The problem is to understand how our ancestors came to be able to construct and interpret quickly and easily all the nested hierarchies, long-distance dependencies, and recursive complexities of everyday speech in any language. Accounts in the wake of Chomsky (Bickerton 1990, 2009; Jackendoff 2002; Hauser, Chomsky, and Fitch 2002) have tended to see the computational brain power required for these tasks as a distinctively human capacity, which came late in our evolution and has no true parallel in any other animal lineages. Such accounts emphasize the evolutionary exceptionalism of human language and reject the idea of a gradual emergence of language from other, nonhuman animal capacities, a position we can call language catastrophism. Balari and Lorenzo instead are gradualists, and they see the powers of the human brain, including those employed in language, as no more than a hypertrophied development of homologous computational capacities found in all vertebrates. Through an analysis of the cognitive requirements of language—from the midst of the Chomskyan wheelhouse, so to speak—they mount strong arguments against the need to invoke uniquely human capacities to explain it.

Despite their gradualism, however, the changing nature of animal computational powers remains fundamental in their account. Balari and Lorenzo recognize distinct types of computation, of graded complexity and with different powers, which give rise in different vertebrate taxa to distinct "computational phenotypes." Relying on the work of language cognitivist Philip Lieberman and others (Lieberman 2006; Okanoya 2002), they propose a common biological substrate for all these phenotypes, a circuit connecting deep-brain structures in the basal ganglia to the cortex. There are two functional components in the circuit: pattern-generating or sequencing capacities in the deep brain and the capacity for memory in the cortex. The first is widespread, probably found in some state of development in all vertebrates, while the second is differentially distributed, absent in some of the simplest vertebrates, modestly manifested in many other, more developed animals, and, in highly elaborated form, very rare. The differences between computational phenotypes depend crucially on differences in the cortical, memory component.

Abstract machines of cognition To understand these differences, Balari and Lorenzo model the functions of Lieberman's deep-brain-to-cortex circuit according to several different "computational automata" of computer science. These automata are *abstract machines*, general schematics that can model the operation and change of dynamic systems. They generalize the patterns of cognition and, in their most complex forms, human cognition.

The simplest computational phenotype can be modeled by a *finite-state automaton*, in which input shifts the machine into one of a determinate number of states. There is no memory involved, but only a response to immediate input. Many animal behaviors, even complex ones, can be modeled by this automaton, and to exemplify it Balari and Lorenzo seize upon the well-studied cases of the songs of Bengalese finches (Okanoya 2002; Okanoya and Yamaguchi 1997) and the alarm calls of Campbell's monkeys (Ouattara, Lemasson, and Zuberbühler 2009). The combinatorial designs of these songs and calls take shape as each motive or gesture in the sequence triggers the next one according to a small set of sequencing rules—algorithms determining what can follow what in the repertory of available motives. Each motive, then, makes possible the shift of the machine into one of a small set of available new states, and the state shift is manifested as the next motive in the song or call. The set of motives and the rules, taken together, constitute a simple formal language (in a logical or mathematical sense), with the motives representing its "alphabet" and the rules its "grammar."

In the next computational phenotype, memory enters the picture. Its operation is modeled by the *pushdown automaton*, in which input enables not only immediate output but also access to a stack of stored data that becomes part of the output. In the simplest version, the stack is "pushed down" like a spring-loaded cart for carrying trays, leaving only the topmost tray available; only the most recently stored data, in other words, alters the operation in response to input. This forms a model of the simplest memory. In more complex versions of this machine—that is, more elaborate kinds of pushdown automata—access can be had to the trays lower down, so that a more varied, longer-term working memory alters the response to input. In a *nested pushdown automaton*, still more complex, the array of stored data is organized such that each entry in a memory stack gives access to its own stack. The primary stack forms a stack of stacks, and the resulting nested hierarchy models the workings of associative memory.

The most complex of Balari and Lorenzo's phenotypes involves this kind of nesting and also one additional machine, the *linear bounded automaton*. This resembles a Turing machine except without its infinite (and in real life unrealizable) "tape," and in its operation the response to input involves reading a whole, discrete set of data, stored on what is envisaged as the bounded, linear tape of the automaton's name. This automaton models a varied, situational memory, in which individual memories are grouped flexibly according to the contexts and situations to which they pertain. It structures output in relation to whole constellations of memories, accessed not only in hierarchized form but also in situated contexts.

Automata and Paleolithic systems

Balari and Lorenzo's primary concern is language processing, and their focus is on the brain functions recruited in it. But now imagine the situation of hominins about 200,000 years ago—incipient *sapiens*, Neandertals, and perhaps some other lineages—whose evolution has furnished them with the cortical capacity that sets in motion operations, in the deep-brain-to-cortex circuit, like those of nested pushdown and linear bounded automata. These ancient humans do not speak fully modern languages, but their communication is eloquent and varied in its indexicality. Their brains do not act only in isolation, like so many unnetworked, organic computers; rather, they *inter*act with other brains, utilizing advanced theory of mind and capacities for observational learning to build cultural traditions of increasing elaboration. How will the new computational capacities shape their cultures, facilitating the creation of systems?

The computational models suggest several answers. In the first place, we might expect that new modes of access to increased memory capacity would enable a deeper *accumulation* of cultural knowledge than was possible without them. A sedimentation of behaviors and ideational patterns transmitted from generation to generation could build up, forming a rich archive of cultural materials passed along as a whole cultural tradition. The growing wealth of materials in this archive would bring about an ordering similar to the hierarchic memory of complex pushdown automata, resulting in a new intricacy of culture and its practices. At the same time, the deepened archive could entail also a new kind of breadth, tapping memory not only in the vertical manner of pushed-down stacks but in an expansive, linear access to situations remembered as a whole. This would of itself create nested hierarchies of memories, as certain memories were constellated together with others in the larger contexts to which they pertained, forming, in effect, stacks of stacks of memories.

A novel *systematization* of thought and action could take shape in such nested hierarchies, with cultural patterns more highly structured than before and passed along in the networked, integrated wholes that now characterized remembered experience. The structure of these systems, also, could form a mediating cognitive layer between stimulus and response, just as the hierarchic or linear memory systems of the most advanced automata interpose structural arrangements between input and output. Such mediation is a kind of *abstraction* of stimulus from response, and such abstraction comes to look like a corollary feature of cultural systems of any considerable complexity. It bears a kinship to another kind of abstraction, one we encountered in chapter 4: the cultural displacement of signs from their immediate contexts that we termed, with Maurice Bloch (2013), transcendental. It is hard to see how Bloch's transcendental sociality could develop very far in the absence of cultural systems, and indeed, the systematic cognitive modes of late hominins were no doubt intimately bound up in every aspect of the release from proximate, face-to-face transactions. In this way, "off-line" cognition would have burgeoned, in which humans came to be able to think not only about the mammoth in front of them but about the one they might confront tomorrow. Off-line thinking is always dependent on systems of working or situational memory.

The hypotheses generated by Balari and Lorenzo's models of cognition are suggestive, for the features of culture they involve were apparent in hominin practices already well before 200,000 years ago. This we know from the systems of Middle Paleolithic technocultures outlined at the beginning of this chapter. Levallois and composite operational sequences reflect cultural archives sedimented to greater depth and showing greater integrated, internal organization than anything involved in Acheulean sequences. They were hierarchically ordered, hence the product of a cognition that could organize in this way the experience of working memory. They abstracted a procedural pattern from its practice to an unprecedented degree. These features of late hominin cultures would become more and more marked, sporadically at first but finally insistently, after the 200,000-year mark. We can summarize them as the *accumulation*, *systematization*, and *abstraction* of cultural archives. Eventually these processes in themselves would serve as the platform on which new kinds of cultural elaboration could arise, with emergent properties of immense consequence.

:3:

Traits of cultural systems

To see how this happened, we must first delineate more fully the features of cultural systems. I described a cultural system earlier as an organized set of signs and behaviors transmitted as a whole across generations. The organized aspect of the system is central—an obvious enough implication of the word *system* itself, but one that needs emphasizing in relation to the bottom-up perspective developed here. What enables us to discern the nascency of systematic culture, which I have placed approximately in the late Lower Paleolithic period, is the appearance of bodies of knowledge and practice that show discrete parts bound in more or less stable interrelations. There is a hierarchic element in the organization of all systems, since they must embody at least a parts-to-whole distinction, but systems of modest or more complexity elaborate this hierarchization to some degree, with internal grouping or nesting of their parts. We have seen such nesting in the distinct kinds of knapping gestures that must be followed in Levallois technique and in the separate operational sequences required to design a composite tool.

Holistic transmission is also central. It reveals the integrated, organized functioning of the array of signs and related behaviors that make up the system—reveals it *as* a system in the unfolding of cultural tradition. At the same time, this holism does not seal off cultural systems from interactions with other systems or aspects of culture, and systemic transmission can never be unchanging. The resulting variation, we will see later, creates evolving lineages of cultural systems instead of static repetition of them.

Cultural systems unite and organize ingredients from the domains that are involved in all culture, systematic or not: semiotic, behavioral, and environmental or material. Culture is semiotic in the way we saw in chapter 4: a transmission of signs that displaces them from the immediacy of their situational usage. Semiosis names the ideational content of the system, though it does not necessarily specify much about how the components of this content are interrelated. (A knapping gesture is a behavior driven by an index, as the knapper acts according to an imagined result toward which the gesture points; the same gesture, taught and learned, displaces or abstracts the sign from its immediate origin, making it a cultural sign.) Cultural systems are made and transmitted, also, in social action (the knapping gesture taught and learned), so the signs that form them are not ideas in some Platonic, inside-the-brain sense but instead entail behaviors. And this behavioral dimension, finally, takes place in an environment, on a lived terrain that anthropologist Tim Ingold (1993, 2000) has conceptualized as a *taskscape*. This is the place where cultural ideas, at whatever level of systematization, are mediated through behavior and social action to meet the material and energetic affordances of the world. Cultural systems arise when all these components are integrated into some kind of fixed and transmissible order (the knapping gesture located in an array of other gestures that only together can attain some desired goal).

Sketches of cultural systems The instances of Levallois and composite-tool technologies provide clear examples of all these features, in their systematic organization, their holistic transmission, and the semiotic, behavioral, and material/energetic ingredients that make them up. Paleolithic kinship systems and systems of social rank must have shown the same features. These existed in groups of hominins well back into the Lower Paleolithic period—the evidence of today's nonhuman primates is enough to suggest a long prehistory for them, in the noncultural form we saw in baboon troops—though we know little about their exact nature or when they took on a systematic form passed down as cultural tradition (Chapais 2008). By about 100,000 years ago, at any rate, artifacts begin to appear in some human groups that were in all likelihood used to mark differences of social position: beads fashioned into bodily ornaments from shells, teeth, and other materials (d'Errico and Vanhaeren 2007). Interactions with the material world created signs of distinction and affiliation. At this moment, we can conclude with modest confidence, the ingredients of a cultural system of social organization were in place.

A similar kind of systematization must have taken hold in protolanguage—or, as I prefer to call it with linguist Jill Bowie (2008), *protodiscourse*. If we accept the inference that humans at about 200,000 years ago did not yet speak a modern language (an inference based on a negative: the absence from their sociality of patterns we associate with language; see chapter 7), we are left to hypothesize about the nature of their communication. This must have involved vocalized calls and bodily gestures of some variety and expressive nuance that functioned as indexes especially of emotional states, prompts to action, and general indications of kinds of action. It seems likely that these were already organized in modest arrays, evincing the presymbolic hyperindexicality discussed in chapter 4 (see also Tomlinson 2015, chap. 3). Protodiscourse at this stage, then, already amounted to a developed communication system passed along in a cultural tradition involving semiosis and its behavioral correlates. If the environmental ingredients seem somewhat attenuated in this example, they would have been clear enough on the ground in communicative negotiations concerning access to sustenance and mates, group hunting or gathering, rearing of the young, preparation of gathering places, and other aspects of social life. These were surely not haphazardly bound to protodiscourse as chance outcomes shaped by its deployment. Rather, we should imagine protodiscursive systems as having arisen from less systematic communicative modes precisely in order to facilitate and develop such important material and social relations.

: 4 :

Emergence

As cultural systems grew more intricate, with greater numbers of components, more complicated relations among them, and clearer hierarchization of parts or processes, a new feature appeared in hominin culture: *emergence*. In the wake of complexity theory and dynamic systems analysis, this concept has been the subject of debates that touch the heart of scientific inquiry. At the same time it has taken on a broad currency in nonscientific academic discourses ranging from quantitative social sciences to computational humanities. It will be an important ingredient in my analysis, so it is well to be clear what I mean by it.

Emergence denotes the presence of properties, features, behaviors, or capacities that appear in systems but are not easily traceable to their component parts. Not all systematic arrays are emergent, but without system, emergence cannot occur. Many of the cases of emergence discussed by theorists involve open thermodynamic systems, which take in energy from and exchange materials with their environments and maintain their identity through networked interactions of their component parts; a living cell is the classic example. Systems need not be open and dynamic in this way to show emergent features, however. Liquidity is often thought of as an emergent feature in aggregates of certain molecules (for example, water molecules), and these aggregates are not in themselves open systems. Hominin culture, meanwhile, has long been able to create static emergent systems, that is, nondynamic systems that show capacities or fulfill functions beyond the reach of their components taken individually; a Neandertal spear is a clear example: an emergent tool.

There is a qualification in my definition—"not *easily* traceable." The question of *how* easy it is to trace the behavior of a system back to the properties of its parts is a fundamental issue in debates on emergence and has prompted whole typologies of it. Philosophers of science William Bechtel and Robert Richardson (2010) frame this problem as one of *decomposability*, in essence asking how full an explanation of a system we can achieve by breaking it down into its parts and understanding individually their properties. This leads them to a tripartite typology. The liquidity of water is well explained through an understanding of the ionic properties of water molecules and the atoms that make them up, so water is a *decomposable* emergent system; the same could be said for the spear. Other systems are *near-decomposable*, and still others appear to be *nondecomposable*. Bechtel and Richardson offer these last systems as full-fledged cases of emergence. Philosopher Mark Bedau (2002), working from artificial-life simulations, arrives at a similar tripartite scheme and calls his categories "nominal," "weak," and "strong" emergence. The emergent feature of life in a cell or organism is one often-adduced instance of the last type, strong or nondecomposable emergence; consciousness or mind in neural networks is another.

But *are* these cases truly nondecomposable? Some philosophers of science think not. They view the whole concept of emergence as an epistemological shadow play, a matter of our state of knowledge and not of any real features of the phenomena in question, and they argue that all putatively emergent systems would, were our knowledge of them complete enough, be fully decomposable to the effects of their components (Kim 1999; for reviews, see Mitchell 2009 and Deacon 2012b). Bechtel and Richardson call this position *epistemic* emergence, contrasting it with *naturalistic* emergence; its advocates tend to be methodological reductivists and to view causality in linear terms. And they might be right, in a world whose workings were transparent, viewed by Laplace's all-knowing demon. Even emergentist Bedau, while elaborating on his category of weak emergence, doubts the relevance to scientific knowledge of the strong type, concluding, "Strong emergence starts where scientific explanation ends" (2002, 11). But the question is not readily resolved, and in the absence of demonic powers an absolute divide between thought and world such as is presumed by the reductivists does not take into account the fact that mind itself is an evolved part of the phenomena it aims to understand, subject in its act of understanding to the conditions and constraints of its own history.

Emergence and feedback A more specific complaint of the reductivists is that the naturalistic position entails a strange view of causality. In order to posit the causes of emergent phenomena, they say, naturalists must imagine a reverse, or "downward," causality, whereby the parts do not cause the system (from the bottom up) but the system, once constituted, mysteriously changes the nature of its parts, in effect causing their properties. Naturalists respond that the reductivists do not consider the networked organization of systems showing emergence. The most important aspects of this organization are the mutual effects of the component parts acting on one another, which is to say the feedback loops connecting them; and feedback looms large in naturalistic accounts of emergence. The molecular components in a metabolic cycle or network, for example, are not transformed into different entities by top-down forces that the system exerts on them; the system as such exerts no force on them at all. Instead, they show new operative capacities by virtue of their connections in the network, entering, in their systemic contexts, into new *functional* relations with the molecules around them (Bechtel and Richardson 2010, xlvi). The altered spatial configurations of enzymes in catalysis and the breaking open of the double helix of DNA in transcription give material form to such new capacities and functions brought about by systemic interrelations. There is no causal mystery involved.

I have had much to say about feedback in earlier chapters and am not finished with the topic. We can look back on the ingredients of niche construction in natural selection—the components of the various diagrams in chapter 3—and see them as equivalents of the molecular components of metabolism, transformed by their functioning within their networks and thereby made the drivers of emergent features of ecosystems, speciation, coevolution, and more. We can also glance ahead to the feedback networks of cultural systems in hominin evolution, which give the components of the systems special, emergent functions. These examples entail not reversed, downward causality but instead nonlinear causality and its consequences. In the naturalistic view, then, emergence refers to the features of a system that arise only through its integrated operation as a system. Its component parts are not transformed into different entities than they would be in isolation but instead take on new functions and operations as a result of the networked interactions they enter into. These interactions typically involve loops of feedback.

Emergence and history

In addition to the reductivists' minimizing of nonlinear causality and its implications for function, there is a second reason to doubt their contention that emergence is no more than an epistemic artifact: history. Accumulated change across time in feedback systems is a powerful generator of nondecomposable emergent qualities, a fact glimpsed already by psychologist Donald T. Campbell in his classic, modestly antireductivist paper of 1974 that coined the phrase "downward causation." Terrence Deacon (2003a) has elaborated this historical dimension in his own naturalistic typology of emergence. For him the most complex forms of emergence—that is, the least decomposable forms—involve two things: the passage of an emergent phenomenon through successive stages such that each new stage is dependent on or constrained by the previous one; and the transmission forward, through the stages of emergence, of replicated structures or systems. Deacon's example of the first of these is the multistage formation of a snowflake as it falls through the air. Each new moment of its design is determined not only by the physics and thermodynamics of crystal-lattice formation and the changing atmospheric conditions it encounters but also by the platform afforded by the structure already formed. In this way it comes to be an emergent "historical record" of the stages of its formation (2003a, 295).

Evolving emergence This historical dimension is compounded by the presence in developing systems of replication, which gives rise to the most complex and least decomposable kind of emergence. In systems that reproduce themselves, states of organization are not merely constrained by previous ones; instead, structures are passed on as a whole to future systems, forming templates for their formation and development. As these are continuously "reentered" in new individuals (transmitted, for example, by the operation of DNA and RNA molecules), lineages of systems emerge, adding a novel kind of open-ended continuity to the sort of history embodied in a snowflake. Deacon proposes that such lineage emergence is characteristic of and limited to living systems. I would rather say, making explicit something left implicit in his proposition: living systems and some cultural systems a few of them have created.

The open-ended histories of emergent lineages would be unchanging and repetitive in a situation of perfect replication and unlimited environmental resources. Given any less-than-perfect fidelity in replication, however, the lineages create an array of varied systems, with each new individual potentially diverging from preceding ones. They now evince two of the ingredients of Darwin's fundamental algorithm, inheritance and variation; and these, as we know, interacting with a situation of environmental constraint, will lead to selection among the variants. In all real-world circumstances, then, not only does the emergence of lineages create open-ended histories but the lineages *evolve*. This describes a general model for the richest emergent history of all, the evolution of life; and here we see from another vantage why it is that (as was noted in chapter 4) living informational systems are always qualitatively different from nonliving ones.

Further abstract machines Darwin's algorithm of inheritance, variation, and selection is nothing other than a grand abstract machine—one of the grandest conceived in modern science. And Deacon's perspective on this machine highlights the historicized emergence of novel individuals or kinds—structures, systems, species, etc.—that it brings forth (see also Pavličev et al. 2016). We can envisage this emergence of new individuals from one more abstract-machinic perspective with systems theorist

Manuel de Landa (1997). He describes a generalized *sorting* machine by which distinct components are ordered across time in different patterns according to the conditions of their formation. His simple physical example is a streambed, where grains and pebbles of different weights and sizes are deposited in layers ordered by the force of the water current moving them. Sorting devices can generate many patterns, involving interlaced meshworks and hierarchic organization of distinct components; de Landa analyzes the formations of sedimentary and igneous rock as, respectively, instances of these two kinds of sorting. Sorted systems, also, record in their structures the history of their formation, and we can see that de Landa's sorting mechanism describes also the formation of Deacon's snowflake.

What happens when the sorting machine works on replicating materials? "A new abstract machine emerges," de Landa writes, "in the form of a blind probe head capable of exploring a space of possible forms" (1997, 263–64). Here again we gravitate toward Darwin's algorithm. The difference in de Landa's abstract mechanics between a mere sorting mechanism and a probe head blindly searching possible forms is equivalent to the difference between Deacon's snowflake and his lineage histories of evolving life. The probe is no real entity but represents instead the dynamic of the feedback loops between the varying forms and environmental possibilities. This dynamic defines *search spaces* of structural or formal possibilities that the probe explores without aim or direction—the *morphospaces* of evolving life-forms, as evolutionists sometimes call them (McGhee 2007).

Cultural emergence

I have dwelled on this abstract mechanism of evolving kinds because its operation came gradually, but finally with decisive impact, to characterize the cultures of late hominins. De Landa's machine took hold in culture when cultural systems reached a certain threshold of complexity and coherence as transmissible wholes; here too it depended on replication. We can retrace the Darwinian logic in cultural terms: if cultural systems are passed on holistically and thus reproduced in their structural arrangement; and if they undergo, in this reproduction, inevitable variation in the material, behavioral, and semiotic components they comprise; and if, finally, they develop and reproduce in environments that enforce constraints of resource and energy; then they will give rise to evolving lineages of systems. Kinds of cultural systems will emerge, persist, vary, and evolve.

Late hominin evolution, in creating advanced cultural systems, created for the first time the conditions in which such emergence could occur in culture. De Landa's probe head began to move through a new kind of search space, not of living systems but of cultural ones. For billions of years the probe had searched the morphospaces of life-forms, but now it came to explore also the spaces of the varying, replicating things some few lifeforms made: their signs and behaviors fashioned into systems. Now there were cultural morphospaces, not only organismal ones.

: 5 :

Functional emergence We must not mistake *what it is that emerges*, an easy error in thinking about emergence. It is not an object of which the components have become invisible or mysterious, as if a Paleolithic spear maker could see no relation between a stone point and the weapon of which it formed a part and was brought up short, at the end of the manufacture, by an unanticipated kind of thing. Instead, what emerges is a behavior or operation of the system overall, based on new functions of its components that could not arise in them individually. A hafted stone point can do things that the same point in isolation cannot do. Because of the new aerodynamic potential it gains from its composite form, it can pierce a hide too tough for it to penetrate when held in the hand. (Extend the aerodynamic potential with an atlatl, thereby creating a more complex composite tool, and you enhance this emergent function.) A seashell is simply a shell, protecting the creature that lives in it. Now involve bead-making humans 70,000–100,000 years ago on the southern coast of Africa (Henshilwood 2007; Henshilwood et al. 2004) or in the eastern Mediterranean (d'Errico and Vanhaeren 2007). In harvesting the shell, perforating it, and stringing it on a strap of grass or hide, they craft a composite implement, like the spear. In hanging this around a neck or wrist or tying it in hair or on clothing, they do something more, which we can speculatively describe: they introduce the composite tool into a system of social differences and affiliations and make of it an insignia. The shell is still a shell, recognizable as such, but it has taken on an emergent function by virtue of its place in a cultural system.

In first approximation, then, what emerges is a novel function—or, more properly, a novel *kind* of function. The strung shell in this scenario, however, is also clearly a sign—an index, to be precise. Does this imply that we should consider semiosis itself, and therefore the animal cultures it founds (see chapter 4), to be emergent phenomena? Yes; emergence as I have described it is an appropriate term for what comes about in the Peircean transaction whereby animal perception connects things in the world, making them into sign and object in regard to one another. Sign, object, and interpretant form a very basic emergent system, one that arises from widely dispersed animal perception and cognition. But semiosis in itself cannot give rise to the most complex emergence, with its histories of evolving cultural lineages. This requires the two additional elements named before. First, the signs must be constellated in organized systems. A gull perceiving a shell in the water as an icon of food has enacted semiosis, but not cultural semiosis and certainly not systematic culture. The shell as insignia, on the other hand, enacts all three: it is a representation of an object (it is a sign), it can be displaced and transmitted (it is a cultural sign), and its meaning emerges as a function of its place in the constellation (it is a sign in systematic culture).

System-wide function

The second requirement for emergent culture is the extension of the system into an evolving lineage of systems through replication and variation. We have viewed this from the generalizing vantage of de Landa's abstract machines, but now we must ground it in ancient hominin experience, where its power to transform culture was far-reaching. The extension to form a lineage compounds through time the new functions that emerge from cultural systems. Since the system as a whole is the unit passed on to form the lineage and the operational unit of cultural tradition, what emerges in this historicized circumstance is not only the novel function of each component in the system—the strung shell become insignia—but also a higher-level function of the system itself, a broad metafunction. Now the system of social positioning, of which strung shells form one part, creates in its entirety its own emergent quality. All of its aspects, including the shells, the social and communicative practices of status and difference, and the interaction of these with material affordances, are rounded into a single functioning whole. The displacement of signs, which we took as definitive of cultural semiosis in chapter 4, is raised to a higher power, and systems themselves are displaced. This kind of cultural transmission has few if any parallels in nonhuman animals.

Historicity and ritual

The rise of such holistic, emergent functioning had broad effects. It introduced the experience of *historicity*, an experience not merely of remembered situations (which many animals have) or even of the cultural memory of a group (which very few have) but of something more: an abstract, systematic lineage of memory—memory displaced, formalized, and repeatable as the system unit that evolves to form the lineage. Here we stand on the verge of organized activities, growing within cultural traditions, that set themselves off from quotidian practice and, through their distinctiveness and repeatable formalism, serve at once as collective retrospection and prospective guides for thought and action. *Ritual* is the best word we can use to name these activities, each of them the in-the-moment manifestation of an evolving cultural lineage functioning to make sense of the past and to frame a desired future. Ritual in this sense is not limited to the kinds of elaborate ceremony we normally associate with the term but embraces more general organizations of social discourse, what anthropologist Michael Silverstein (1993, 2003) has termed *metapragmatics*: durable, overarching systems of indexes shaping discourse and action. The indexical connection is important, and from our deep-historical vantage we can see that the metapragmatics embodied in ritual represents a consolidation of the systematized indexes that came early to hominin cultures. The first rituals depended on such hyperindexicality but did not require full-fledged symbolism, and I will propose in chapter 7 that they developed already in a presymbolic world, before modern language had coalesced: the world of protodiscourse. Perhaps the teaching of techniques on the taskscape, transmitting systems such as the Levallois operational sequence, formed some of the earliest of these indexical rituals.

Systemic abstraction A related effect of the emergent function of whole cultural systems was to mark them off and underscore their integrated coherence as systems. This *abstraction* was, we have already seen, the equivalent at the systemic level of the abstraction of signs that I described in chapter 4. The loosely knit systematicity of Acheulean operational sequences was superseded already in the Middle Paleolithic era by the more marked systems of Levallois technologies; in its turn, the boundedness and integration of Levallois systematicity pale in comparison to those of the shell-necklace system I described above. As the borders of cultural systems were more clearly marked, so were their internal structures; abstraction fostered consolidation. The advanced computational cognition of late hominins no doubt played an important role here, since systematic structure in culture is bound up with capacities to distinguish parts of wholes and organize them into hierarchies. This cognition probably evolved in a feedback between sociality on the taskscape and wetware in the head: an instance, in other words, of cultural niche construction altering the genotype. This outlines a "social intelligence hypothesis," in other words, of a kind seeking to explain the mutual formation of modern minds and cultural systems (for other kinds, see Dunbar 1998; Holekamp 2007; and Cheney and Seyfarth 2007). It is never enough, in contemplating the rise of advanced computational capacities, to remain inside the skull.

Structural stability The consolidation of the internal structures of cultural systems served to strengthen and stabilize them. The Paleolithic systems that archaeologists enable us to discern are often robust ones, maintained and replicated across staggeringly long *durées*. Among technocultural systems, Levallois techniques show a consistent life lasting 200 millennia, and Acheulean bifaces, while manifesting a less complex system, persisted without wholesale change across a million years and were fashioned by a number of different hominin species. The bead making I have referred to, creating small ornaments for stringing and hanging, is more recent than these, but it was a widely dispersed industry, stable in its systemic outlines, and it recurs in the archaeological record across the last 100 millennia from southern Africa to Eurasia and beyond. It is hard to imagine functions for these beads not involving social marking of some sort, and therefore hard to imagine that they did not in each local recurrence form elements in cultural systems of considerable complexity. These examples of *systemic stability* can be multiplied: incised ostrich eggshells, fragments of what were probably water containers bearing marks of ownership or social affiliation, reach back 60,000–70,000 years in Africa and are little different from those made by San and other peoples within the span of ethnographic memory—even to the point of similar etched patterns (Texier et al. 2010). Flutes crafted from birdwing bones and of remarkably similar design survive from more than 40,000 years ago in Germany (Conard, Malina, and Münzel 2009), from 20,000–25,000 years ago in France (Buisson 1990; Morley 2013), and from 8,000 years ago in China (Zhang et al. 1999). Even if we resist anachronistic scenes of fully modern musicking, speculation about the uses of these instruments points toward highly developed, broadly similar cultural systems (Tomlinson 2015).

Continuity versus independence

Such consistencies attest the robustness of the cultural systems involved but not the continuity of particular traditions or the diffusion of specific cultural practices. This seems paradoxical, and smoothing out the conceptual knot will reveal another important feature of the emergent functions of the systems. Local cultural traditions certainly existed in all early human groups, and some were no doubt long lasting. Now and then Paleolithic archaeological evidence seems plentiful and precise enough to bear witness to this continuity, as in the case of the extraordinary traditions of flute making and miniature animal carving unearthed in recent decades in a number of neighboring sites in southwest Germany (Conard et al. 2015). Far more often, however, the evidence, fragmentary and dispersed across large geographical and/or temporal spans, does not grant adequate resolution for us to assign separate sites, or even separate strata at the same site, to the same local culture. So, while the view from a distance that archaeology affords can foster the illusion of a continuity of individual traditions across early human culture, the reality was probably very different: much more diverse and scattered, with small groups moving across the landscape, following shifting resources in response to the volatile, quickly changing climate of the Upper Pleistocene age, and independently inventing similar cultural systems to answer their needs (see chapter 6).

The robustness of the cultural systems cited above, then, gives the appearance of continuities across all instances of the systems, but this is probably most often a false impression left by the low resolution of archaeological data. The constructive similarities of the German and French flutes, for example, have encouraged speculation about a connected, European growth of musical technology, but the chronological distance between them, approximately 20,000 years, is too vast for an unbroken line of cultural transmission to be plausible. In fact, the humans who made the earlier instruments may have died out or been driven out of Europe by periods of climate deterioration before different human groups repopulated the region and produced the later flutes (Bradtmöller et al. 2010). If we include the Chinese flutes, the geographical and temporal distances become still more daunting for hypotheses of connected cultural traditions. And in the other cases named above, scales of time and distance expand dramatically: 100 millennia and two large continents for bead making, 200 millennia and two continents for Levallois systems, a million years and wide dispersion for Acheulean ones—not to mention several different species employing them. Here there should be no illusion of continuous traditions.

Convergence and attendancy

For cases such as these of robust and similar systems dispersed widely in time and space, we need a model that does not rely on hypotheses of connected traditions of vast duration. The systems must instead have tended independently to assume similar structural outlines in the presence of similar needs, capacities, behaviors, and environmental affordance. In the cultural morphospaces where they took shape, we can think of such systems as *attractors*, states of relative stability arising from the dynamic interaction of similar conditions and resources. They brought about a convergent cultural evolution, and elsewhere I have termed the forces that militated for it the *attendancy* of a cultural system on the constellation of elements from which it arose (Tomlinson 2015). This term makes a bow to the French term *tendance*, used by archaeologist André Leroi-Gourhan to refer to the "canalization" of technique evident in recurrent forms such as the Acheulean biface (Leroi-Gourhan 1993; Gamble 2007, 221), with the difference that it is the system in relation to culture and environmental affordance that I wish to highlight, not the consistency of resulting form. (Philosophers might prefer the term *supervenience*, but I avoid it because it connotes an ontological state more than a tendency or directedness of process and relation.)

Bead making again provides a useful example. Once social organization had reached a certain level of intricacy, it began to form stable intragroup distinctions of rank or affiliation. In the presence, then, of long-used techniques of shaping, piercing, and drilling materials of various sorts and fashioning straps of hide or grass, it needed only a small, near-inevitable swerve in technological traditions to make insignias of social difference. No doubt, indeed, the move was also made in more perishable ways, through body painting and scarification, clothing, and the like. Similar bead-making systems, and with them similar emergent functions for materials long used in other ways, sprang up over and over again because of the tendency of the parts of the system to fall into *this* determinate functional relation rather than other ones: to fall toward *this* attractor in the cultural morphospace. The same kind of attendancy can also explain the similarities of flutes from Germany and China, separated by 30,000 years and 5,000 miles. Take a human with a deepening ritual culture, armed with the basic cognitive means that underlie music still today (Tomlinson 2015), and numbering among its hunted prey large, hollow-boned birds, and this musical tool formed a deep cultural attractor, making convergence on it highly likely. It is probably only a matter of time and good fortune before archaeologists uncover more flutes of similar design at other Paleolithic sites, unrelated to those in Germany, France, or China. In the same way drums, bull-roarers, and musical bows probably arose in many times and places without cultural transmission or diffusion as such.

Attendancy, then, is not a model of cultural transmission but of convergent propensities. This differentiates it from anthropologist Dan Sperber's (1996) appeal to the attractor model to explain the diffusion of similar cultural phenomena. Sperber, moreover, also attributes his similarities to a Fodorian modularity of mind that leads humans toward certain mental representations more readily than others, and this too is not to my point—or rather, it overemphasizes the in-the-brain aspect of a phenomenon that gains much of its force outside the brain. Attendancy, instead, whatever its reliance on deep commonalities of human minds, modular or otherwise, exerts its force in structures built beyond minds themselves, in the space of the social and environmental negotiations that give rise to systematicity. The robustness of Paleolithic cultural systems relied in part on local traditions of cultural transmission, but it relied also on their attendancy on general alignments of brains, bodies, taskscapes, and ecosystems. The result across deep history was frequent iterations of the systems among different human groups widely dispersed in time and place.

: 6 :

This chapter has followed the developments that set early human **Autonomy** cultures on a course diverging from all other animal cultures: the rise of cultural systems in accumulating cultural archives, the emergence within them of novel functions, and their formation into evolving lineages. From these arose a new kind of system-wide function, and this in turn gave rise to several features unique (at least in degree) to human cultural systems: their attendancy on conditions of knowledge, behavior, and resource, with the resulting independent formation of similar systems in similar conditions; the abstraction of the systems from other aspects of culture, leading to the consolidation and robust durability of their structures; and the ritual historicity of the lineages. These features seem to have appeared already in the late Lower Paleolithic era, and they are highly developed by the Middle to Upper Paleolithic (or, in African terminology, Middle to Later Stone Age) horizon.

Together these features conduced to a final, important novelty of Paleolithic cultural systems: their growing *autonomy* from the biocultural dynamic that spawned them. As the systems were abstracted and consolidated, their components, linked in the network of relations that gave them their functions, came to reinforce the structure of which they were a part. A self-organizing, reciprocally strengthening aspect arose in these systems, further consolidating their structures and the mutually determined functions of their components. Once a tradition of flute making was established in a particular human group, for example, a resource such as hollow bird-wing bones was channeled into the system of ritual musicking. At the same time, the availability of such bones directed and structured the nature of the ritual, right down to the soundscape it shaped. The resources and behaviors pulled each other into a reciprocally defining interaction. Similarly, seashells or teeth from hunted animals came to organize social-marking systems even as they were exploited in them, their cultural function transformed. Or again: ocher was likely at first a haphazardly (if frequently) used resource, but later, once body painting came to be linked to social difference, it probably was harvested for more determinate purposes, valued for its role in achieving them, and formative of the marking system itself. We can imagine resources, behaviors, and ideas alike coming under pressures exerted by all the consolidated systems of which they formed a part, pressures arising from the fact that the larger wholes were experienced as present-time instantiations of historicized lineages imbued with valued cultural function.

Cycles and epicycles In this way the developments following on system-wide function and militating for systemic consolidation opened a new distance between the system and the society that created it, the distance that I have termed *autonomy*. This needs to be seen in its broad evolutionary context. The lineages of systems arose, at the most general level, from a biocultural dynamic of evolving animals in changing niches—from, that is, the feedback cycles of cultural niche construction examined in chapter 3. These, however, finally produced, from within the advancing cultural systems, phenomena whose emergent qualities set them dramatically apart: novel kinds of systems. These were new in kind because they stood outside the cycles as *epicycles*, shielded in some degree from the turning and change of feedback. Their partial autonomy gained them a robust and stable structure and hence a more forceful role in the taskscapes of their creators than less stable cultural systems. Because of this they came to approximate systems of guidance or control, channeling the more fluid feedback dynamics of culture according to their own stable arrangements.

At the end of chapter 2 I introduced elements from cybernetic theory and the schematics of feedback systems that resemble these epicycles: the external controls guiding systems through *feedforward*. Indeed, a general definition of an epicycle can be advanced in just these control-theory terms: *a cultural epicycle is a network of systematized elements that emerges from evolving feedback cycles and assumes a structure so integrated, autonomous, and durable that it comes to exert a control-like function over the cultural cycles from which it arose*. Epicycle, then, names the product of a particular kind of emergence, indeed the ne plus ultra of emergent functionality, in which feedback-driven elements become so stably systematic that they act locally like feedforward ones. Of all the novelties that the systematization of late Paleolithic culture introduced, cultural epicycles were the most remarkable—and arguably the most powerful. Culture engendered systems of sufficiently consolidated integrity to direct cultural practice from a place somewhere outside it, even beyond the control of the culture-producing humans involved. For thousands of millions of years before this, genes had exerted a strong if general force channeling the growth and lifeways of organisms. Now, for the first time, one life-form, or several closely related life-forms, created a nongenetic phenomenon that could exert a kindred force.

Through what mechanisms could the independence of the epicycles arise? We will see in chapter 7 that an important ingredient in their formation is the existence of broad prior conditions, possibilities, or constraints. Whatever the elements involved in an emerging epicycle, as their systematization takes shape they push up against conditions that were not encountered in the elements individually or in their more loosely ordered state. It is as if these conditions are activated at a certain moment of systematization, their effects now felt where they had not been felt before, their force now recruited in the emergent dynamic. The conditions come then to play a part in systematizing the elements, helping to fix the emerging epicycle in one morphospace rather than another and to determine its control functions as a whole in the broader feedback networks from which it originated.

The elements involved in epicycles can no doubt take many forms; the concept concerns a type of dynamic system and does not specify the ingredients it contains. In this book I will discuss only epicycles that arose from late hominin cultural systems, which involved in all cases the semiotic, behavioral, and material ingredients involved in the making of taskscapes. Such cultural epicycles are probably unique to the hominin line, emerging as they do from complex systems of a sort occurring rarely if ever outside it. It seems probable, however, that epicycles of noncultural kinds have arisen in evolved systems of all sorts and at all scales throughout the history of life, and that the concept has an application generalizable far beyond culture. In the microscopic direction, epicycles might help us to understand systematic phenomena reaching all the way to the molecular metabolic networks whose durability and control functions have been advanced as the foundation of homologies extending across the living realm (Wagner 2014; Pavličev et al. 2016). And in the macroscopic direction, epicycles might clarify the broadest control mechanisms of ecosystems and even of the biosphere as a whole.

CHAPTER 6

THE FIRST 150,000 YEARS OF *HOMO SAPIENS*

:1:

Understanding complex histories is like digging in loose sand: each deepening of insight brings a tumble of new questions all around. So it is with the difficult historical questions scientists aim to answer. The origin of the cosmos, the rise of earthly life, and the emergences of animal consciousness and mind are all topics in which growing knowledge in recent decades has triggered sand-slides of new puzzles. The emergence of modern humanity is another such topic. Here huge recent gains in various fields—signally archaeology, paleoanthropology, climate history, and genetics—have led to something approaching consensus on many issues, but each consensus seems to steepen the grade down which questions spill from all sides.

Out of Africa, many times

The foremost recent consensus involves the question of where the humans who today populate the world came from. *Homo sapiens*, we have learned from genetic, fossil, and archaeological evidence, took its modern form about 200,000 years ago in Africa (Stringer 2007; Klein 2009), after a little-understood nascency that reaches back at least another 100,000 years (Hublin et al. 2017; Richter et al. 2017). Our origin in Africa is now so settled a matter that it is difficult to remember that it was the subject of heated debate as recently as the 1990s, when it competed with theories of "multiregional" origins proposing that modern humans evolved independently out of several earlier hominin populations in Asia and Europe as well as Africa. The non-African populations that featured in those theories were real, though their subsequent evolution into modern humans is now not accepted. They also had spread from Africa, but about 1.7 million years ago, long before *Homo sapiens* existed. In this early form they are known as *Homo erectus*, a branch of the African *Homo ergaster*, the first large-brained hominin, and theirs was the earliest movement of hominins out of Africa that we know of. Other hominin exoduses from Africa followed. There must have been a split, for example, in the common ancestor we share with Neandertals, usually identified as *Homo heidelbergensis*. It led to the independent evolutions of Eurasian Neandertals and African *sapiens* and likely took place sometime between 700,000 and 400,000 years ago (Endicott, Ho, and Stringer 2010). And even an early form of *Homo sapiens* may have migrated out of Africa and interbred with Neandertals more than 270,000 years ago, genetic analysis has suggested (Posth et al. 2017).

Today's humanity, however, does not stem from these early expansions but from a much more recent one. The story is a complicated one, glimpsed only in fragments. The pace of sapient migrations out of Africa picked up by about 130,000 years ago, and archaeologists have uncovered traces of their habitation of the Levant that might stretch across the next 40,000 years or so. There may have been many such movements, perhaps abetted by changes for the better in the climate of northeast Africa, but none of them seems to have resulted in permanent colonization; indeed, across many millennia Neandertals inhabited some areas *after* sapient humans had been there (Shea 2007). Finally, sometime after 70,000 years ago, a single lineage of *Homo sapiens* in Africa—most likely central East Africa—expanded permanently beyond the continent; it is the founding lineage for all humans in the world today. For this population there is genetic evidence of two primary routes out of Africa. One traced a course northward through the Levant and ultimately into central Eurasia and Europe, while another one, far to the south and perhaps somewhat earlier, moved east along the southern coast of Asia, reaching India, China, Southeast Asia, and Australia (Atkinson, Gray, and Drummond 2008; Beyin 2011).

Each new movement of human populations we discover raises new questions: What caused the collapses of the early incursions of *Homo sapiens* into southwest Asia? How were they replaced by Neandertals, when, and why? Given the two main migratory routes of the founding lineage, did these populations divide in Africa or later, after they had reached southwest Asia? And here is one more question, to which some answers have recently been offered: As the founding lineage expanded and migrated, what was its relation to the older, indigenous hominin populations ("archaic" populations, as they are usually called) that it encountered along the way? Evidence of such encounters has been tracked in genetic variants in present-day non-African humans, which indicate some interbreeding with Neandertals and with the mysterious Denisovans of southern Siberia—the only two extinct kinds of humans whose genomes have been sequenced, thus allowing comparison with ours (Green et al. 2010; Reich et al. 2010). (At present, the prospects are not good for sequencing the genomes of older taxa like *ergaster* or *heidelbergensis*.) There is even evidence of ancient introgression of Neandertal genetic variants *back* into populations in Africa, indicating two-way gene flows between Eurasia and Africa during the era of worldwide expansion (Wang et al. 2013).

Moreover, the founding lineage itself did not migrate only out of Africa but also expanded back through it. The move into Eurasia, it seems likely, should be viewed as a late aspect of a growth in this lineage that had already been under way for thousands of years inside Africa (Atkinson, Gray, and Drummond 2009). This is part of a shadowy history of ancient sapient populations within Africa, the complexities of which genetic modeling has recently begun to illuminate: a history of groups dividing from one another, later remeeting and interbreeding, and even mixing along the way with nonsapient, archaic humans, just as they eventually did outside Africa (Forster 2004; Hammer et al. 2011; Schiffels and Durbin 2014; Hsieh et al. 2016). The movements of populations of *Homo sapiens*, in short, have been many, and another recent consensus affirms that this deep history can be reconstructed only by taking into account the meetings, separations, expansions, bottlenecks, and migrations that geneticists can now track through both present-day and ancient DNA (Hunley, Healy, and Long 2009).

Climate concerns

We know little about the immediate preexpansion history of our founding lineage in Africa. Genetic evidence, here again indispensable, suggests that its numbers were small, with its population even dwindling to a few thousand mating pairs at the low point. Several causes of this bottleneck have been proposed. They usually feature climate deterioration, and some emphasize the eruption of the supervolcano Mount Toba in Sumatra about 74,000 years ago, supposed to have caused a long "volcanic winter" and even to have initiated a global climatic downturn (Ambrose 1998; Lahr and Foley 1998; Mellars 2006b). These hypotheses are debated in their specifics, but they signal another broad consensus that has grown up in the last few decades: as we track the movements of human populations and the evolution of hominins in general, we must take account of the fact that they were channeled and constrained by the dramatic instability of climate throughout the Pleistocene (Finlayson 2005; Potts 2013; Richerson and Boyd 2013; Timmermann and Friedrich 2016; deMenocal and Stringer 2016).

The instability was redoubled in the Upper Pleistocene epoch, which reaches from 127,000 years ago down to the beginning of the Holocene, 11,700 years ago. The Upper Pleistocene began with a long, warm interglacial period—perhaps a stimulus for those early, impermanent movements of *sapiens* into the Levant—but mostly it was a time of cool or cold climates. The cooling period, especially marked from 75,000 years ago, was characterized also by abrupt fluctuations—some as fast as a human generation—with the coldest subperiods (stadials) alternating with subperiods of relative warmth (interstadials). These climate swings, moreover, did not have uniform effects across different regions. A new glaciation in Europe, for example, coincided with climate deterioration in other regions as well, but the downturn took one form in the Levant, another in northwest Africa, and still another in southern Africa—not to mention Asia and Australia (Finlayson and Carrión 2007; Shea 2007; Müller et al. 2011). The southerly effects did not involve glaciation but tended to include drought and the expansion of deserts. Then as now, the global weather system was a single system with myriad local effects. No attempt to understand the movements of ancient human populations can afford to ignore the record of these effects, which paleoclimatologists are constructing with ever-greater precision.

In the midst of the long cooling, a series of benign interstadials set in about 60,000 years ago, and another point of growing agreement about ancient human movements assigns the expansion of the founding lineage to this period of warm, wet climate, and especially to a long interstadial from 55,000 to 49,000 years ago. Some have even tried to match specific migrations to individual, brief climatic swings, an alluring specificity bound to remain speculative, given the low resolution that archaeological evidence and dating methods yield for human movements in this distant period (Müller et al. 2011). The general timing of the benign period, at any rate, accords well with several kinds of evidence for *Homo sapiens*' expansion through Europe and western Asia. The earliest artifacts in Western Europe associated with *sapiens*—associated by inference from the kinds of artifacts involved, which are unlike Neandertal artifacts, but not by the copresence of human fossils—reach back more than 40,000 years (Higham et al. 2012; Mellars 2006a). Positing exoduses from Africa 50,000–60,000 years ago, with populations reaching distant points on the order of 5–15 millennia later, makes sense. But the picture is certainly not clear in all its aspects. The move of the founding lineage through southern Asia to Australia, in particular, suggests a somewhat earlier migration, since the most recent, securely dated evidence points to *sapiens* in Australia between 59,000 and 65,000 years ago (Clarkson et al. 2017).

Tracking haplogroups Genetic analysis supports in a general way this timeline for the expansion of our ancestral lineage. It relies on the tracking of specific mutations in modern human groups to reconstruct a migratory history of population movements. Because such evidence is readily converted into misleading headlines, and because I have already referred to it more than once, it is worth pausing to specify how it works (Cavalli-Sforza and Feldman 2003; Stone and Lurquin 2007).

To track the expansion of our founding lineage through the world, paleo-geneticists start from the basic assumption that it divided as it colonized new areas, part of it staying in place and another part moving on. As the subpopulations separated across the landscape into distinct breeding groups, their individual, collective genomes accrued random mutations in the way all genomes do. Most of these were "silent" mutations, without effects on the phenotype and hence not subject to natural selection; they were, in other words, the product of genetic "drift." A repeating pattern was formed, then, in which each part of the lineage migrating into a new territory—a "founder" population—carried with it the mutations of the population from which it had separated and then introduced new ones of its own. Across successive divisions and migrations a "serial founder" pattern was built, in effect a genetic record of the successive founding populations, each inhabiting a new area.

The record of these accrued mutations is usually read by analyzing, in distinct modern populations, mutations in DNA varieties that do not recombine in the process of sexual reproduction, since such nonrecombinant DNA accumulates and passes on every silent mutation that occurs, forming complete records of drift. (Other, more complex methods of analysis have recently been devised for recombinant DNA.) Two sources of nonrecombinant DNA have featured in this work: DNA found in cell mitochondria, which is passed almost exclusively from mother to offspring, and Y-chromosomal DNA passed in male gametes, or sperm. Gamete cells have one copy of genes rather than the two copies found in nonsexual human cells; they are, that is, *haploid* rather than *diploid* cells, and from this fact comes the name for the distinct groups of mutations used as markers and the distinct populations that carry them: "haplotypes" or "haplogroups."

The migratory history read in haplogroups can be followed from the founding lineage all the way to the relatively recent human colonizations of North and South America and the Pacific islands. There is much fascinating information here: for example, the suggestion of the southern route out of Africa and along the Asian coastline, mentioned above, by a population that separated early on—even within Africa—from the groups that took a more northerly route to western Eurasia. (Perhaps the puzzle of early sapient habitation of Australia would be explained by finer-grained timings of these separate migrations than we now possess.) To assign calendar-year dates to the haplogroup movements, however, requires us to calibrate the timing of genetic change, and this involves more assumptions: that mutational drift occurs at a stochastically constant rate, yielding a genetic "clock"; and that the conversion of this regularity into a timeline can rely on gauging it against some date fixed by paleontological or archaeological evidence for the advent of genomes known to us—for example, the date of the split that led to modern human and chimpanzee genomes or of the first appearance of humans in the locale of a particular haplogroup. Given the approximate nature of such dates (datings of the chimpanzee/human division range between 5 and 8 million years ago), not to mention the practical difficulties of counting mutations and sampling appropriate populations, the calendar-year timeline for human haplogroups is at best inexact. Nevertheless, a recent calibration places the *terminus post quem* for the move out of Africa between 62,000 and 95,000 years ago—a range that matches in a general way when archaeologists would expect it (Fu et al. 2013).

Population histories A further consensus is evident in this research on genetics, founding populations, and migrations: as we reconstruct our deep history, it is no longer enough for us to speak of *Homo sapiens* as a whole, on the one hand, or to extrapolate only from the evidence of single sites, on the other. The unit of study for Paleolithic humans has changed somewhat, to become whole populations reconstructed in local ecologies through evidence from several fields (Shennan 2001, 2002; Steele and Shennan 2009; Powell, Shennan, and Thomas 2009; Richerson, Boyd, and Bettinger 2009). Fine-grained archaeology and paleontology remain foundational, needless to say, but their findings are joined with those of paleogenetics, paleodemography, and paleoclimatology to gain a picture of regional and subregional populations and their behaviors. Thus, even as genetic evidence demonstrates that we are all descendants from a single African lineage, it paints a picture of incessant movement that divided and merged groups of early humans, created both long-term and transient settlements, and resulted in some groups thriving while others dwindled. The movement carried individual sapient populations into new ecologies—or perhaps in many cases the humans were carried along as component parts of ecosystems slowly shifting with changing climates. In the former case, the new challenges of a new niche were quickly felt, while in the latter they might arise only when an abrupt turn in the climate occurred. Thus, the repeated flux of sapients into and out of the Levant around 100,000 years ago may have followed the rhythm of warm, wet periods, which made the area into an extension of the northeast African ecology, alternating with cold, dry periods, which pushed the humans back into Africa and brought Neandertals down from the now-uninhabitable north (Shea 2007). Other, similar cases have been proposed: a climate-driven collapse of local sapient populations and cultures is now proposed for Europe about 35,000 years ago (Bradtmöller et al. 2010), and yet another one may have occurred about 70,000 years ago in southern Africa—though here precise paleoclimatic information is lacking (Jacobs et al. 2008).

: 2 :

The consensus views considered so far are both topical and methodological. *Homo sapiens* arose not in all the regions of Eurasia that earlier hominins had inhabited but only in Africa, taking its modern form there about 200,000 years ago. Close to 150,000 years passed before sapients gained a permanent foothold anywhere outside Africa, but the moment when they did so marked an important event, in which a single lineage expanded fast and far, both in Africa and out, with some interbreeding with the indigenous humans they encountered. The lineage's movements can be followed through genetic analysis as well as fossil and archaeological evidence. Everywhere humans went, their movements depended on the vicissitudes of the unstable Upper Pleistocene climate, which grew especially cold and erratic starting about 75,000 years ago.

Revolution and counterrevolution

Along with all these advances in our knowledge has come a major puzzle, the largest sand-slide of all. What happened to *Homo sapiens* between the appearance of its modern form in Africa and its permanent move into other continents? Although we still lack, for this crucial period in Africa (referred to as the Middle Stone Age), anything approaching the depth of knowledge we have for the same period in Europe (the late Middle Paleolithic), intensified archaeological and paleontological efforts in recent decades have begun to fill in the picture. Researchers active in many different parts of Africa have shed light on the behavior of sapients there, through careful amassing of evidence as well as some stunning, headline-grabbing finds. The first major impact of this work has been to lay to rest the hoariest archaeological narrative about human prehistory: the story of the European Upper Paleolithic "revolution."

It is an old, familiar story. At the close of the Middle Paleolithic period, modern humans swept across Europe, pushing disadvantaged, brutish Neandertals ahead of them toward their last refuge on the Iberian Peninsula and doing things Neandertals couldn't do: covering cave walls with paintings, carving Venus figurines, and fashioning an array of new tools and weapons. Where these moderns came from was not entirely clear, but it was somewhere to the east, probably in central Eurasia. Whatever their origins, they brought a new, tall, gracile look, a nobly smooth and high forehead, and an advanced cognition and behavior nowhere else known on the planet at the time. They invented art, music, myth, religion, and more. Human modernity began with them, a European creation.

The problems with this story start with the mysterious origins of these uniquely capable moderns, and the evidence that has formed the out-of-Africa consensus has resolved the mystery in a way that early archaeological Eurocentrists did not often predict. Today's humans of Western European ancestry represent haplogroups descended from the same African lineage that peopled the rest of the world, and the progress of those ancestors can be followed, in a general way bolstered by some fossil and artifactual finds, from their point of entry in southwest Asia (Bar-Yosef 2007).

The progress was, however, neither consistent nor inevitable, and this is the next problem with the old story. There was no steady, triumphal advance across Europe, but instead local movements of small populations of *sapiens*, expanding in warm periods and suffering decline or even extinction in cold ones. The makers of bird-bone flutes and exquisite, mammoth-ivory figurines in the upper Danube valley more than 40,000 years ago, members of what archaeologists term Aurignacian cultures, may have died off entirely when a new climatic deterioration set in, to be replaced only several millennia later by different sapient populations that developed subsequent, Gravettian cultures; at the very least, the Aurignacians had to abandon these areas, driven to more temperate refugia as times grew harder (Bradtmöller et al. 2010). For Aurignacians and Gravettians both—and European humans long after them, for that matter—there is no convincing case for continuous, progressive "conquest" of new territory. Fluxes of advance and retreat, expansion and decline, like those 50,000 years earlier in the Levant, were probably the rule across most of Europe at least through the Last Glacial Maximum, which began slowly to loosen its grip only about 18,000 years ago.

Giving Neandertals their due

Another problem: in order to make its case for behavioral innovation and steady migratory progress, the story of the Upper Paleolithic revolution much underrated Neandertals (Langley, Clarkson, and Ulm 2008; Nowell 2010). The more we learn about our cousins, the more extraordinary we find their lifeways. Neandertals ranged across Eurasia from southern Siberia and the Levant to Portugal, they withstood environmental hardships such as no humans had faced before them—their stocky frames and short limbs were adaptations for cold weather—and they relied for their subsistence at least in part on bringing down the largest animals in their biomes (and outcompeting ferocious carnivores for them). Evidence for their interbreeding with *sapiens* underscores how close they must have been to us, as does archaeological reconstruction of their varied and proficient technologies and—more controversial but by now widely accepted—traces of personal ornaments and burials of the dead. There is even evidence of Neandertals etching abstract designs and building arrangements of stalagmites deep in caves (Rodríguez-Vidal et al. 2014; Jaubert et al. 2016). Their exact cognitive differences from us remain the subject of intense speculation and debate, with one well-developed hypothesis, for example, proposing as the key that their working memory system operated in a way subtly different from ours (Wynn and Coolidge 2004), and another, which we encountered in chapter 3, hypothesizing a specific, Neandertal modularity of mind, likewise different from ours (Mithen 1996). But for those who accept that there were general cognitive differences at all, they are today usually considered to be small ones, and speculation is now more prone than not to assimilate Neandertal behavior to ours. The cause of their extinction remains mysterious, although most views maintain that the introduction of sapients into their territories by 40,000 years ago had something to do with it. Even if this is true, however, it warrants no assumption of large differences between their capacities and ours. Recursion equations of the sort discussed in chapter 3 have been run to model situations of sapient-Neandertal competition, and they show that the smallest of advantages on the sapient side could have worked large changes in Neandertal viability in a span of remarkably few generations (Zubrow 1989). These effects could have been magnified by demographic circumstances if, as some evidence suggests, Neandertal groups in late Ice Age Eurasia were small and isolated from one another.

The challenge from Africa

The weaknesses of the narrative of the Upper Paleolithic revolution in Europe are several, then, even if we focus only on the situation there. If we widen our view to include Africa, it becomes clear that the most glaring weakness of all is the absence of an account of the nature of earlier *Homo sapiens*. This is where recent advances in African archaeology weigh in with particular force. These have included some much-publicized discoveries, most famous of all a series of finds along the South African coast, where a single site, Blombos Cave, has yielded bone tools, shells perforated for stringing, and etched pieces of ocher all more than 70,000 years old, as well as abalone-shell "paintpots," with pigment and stirrers still in them, dating back another 30,000 years (Henshilwood et al. 2002, 2004, 2011). The impact of recent African archaeology, however, is not just a question of such arresting artifacts; finds from many parts of the continent have been reshaping our views with their sophistication, variety, and antiquity.

The turning point in the recognition of this new perspective came in 2000, when Sally McBrearty and Alison Brooks, two archaeologists long active in Africa, published a synthetic article entitled "The Revolution That Wasn't: A New Interpretation of the Origin of Modern Human Behavior," an Afrocentrist tract that ranked for the next decade among the most cited articles in Paleolithic archaeology. For well over a hundred pages, McBrearty and Brooks surveyed African developments stretching from the end of the Early Stone Age to the Later Stone Age. They tabulated the hominin fossil remains, analyzed the artifactual record, built timelines for the appearances of a wide assortment of African technologies, speculated on early human social practices, and marshaled evidence that many behaviors thought to have been European innovations in the Upper Paleolithic period occurred much earlier in Africa.

Behavioral gradualism McBrearty and Brooks (2000, 529) see human modernity as the result of a "gradual assembling" of behaviors and traits over a long period reaching back to *Homo sapiens*' immediate African predecessors 300,000 or more years ago. This assembly involved mostly a set of cultural events, and it occurred variably across many small populations in many different areas, without revolutionary turning points along the way. The main shifts in cognitive capacity that enabled the process occurred at the beginning of the period, and at its end the shift to the Later Stone Age was caused by population growth and the new interchanges and resource pressures it brought about after 50,000 years ago. The key concepts here—*gradualism* in historical change, piecemeal *accretion* of technologies, and *variability* of behavior—were not new in models of our prehistory, but McBrearty and Brooks joined them together in an African context and, in doing so, forged an archaeological consensus that accorded with the new biological consensus on our origins.

An African revolution? However, as with all the other points of agreement I have taken up so far, the sand began to slip, and new questions and debates were quick to arise. In the first place, some scholars were loath to relinquish revolutionary models, and they took on board the new Afrocentrism only by transferring to Africa the model of a sudden appearance of modern humans out of nonmodern ones (Mellars 2006b, 2007). For evidence in this effort they turned especially to recent archaeological discoveries from the southern end of the continent. These defined two Stone Age industries that succeeded one another between about 75,000 and 60,000 years ago, termed the Still Bay and Howiesons Poort industries after their type sites. (Different stratigraphic layers at Blombos Cave have yielded artifacts from both.) The industries featured technologies that seemed not to be amply attested before them, and these resembled the markers of the supposed European revolution, if in their own local versions: long, narrow stone blades; tools of several sorts fashioned from bone rather than stone; incised pieces of ocher and ostrich eggshell; marine shells fashioned to make personal ornaments; and more. Here, the revolutionists argued, was an explosion of technological innovation, social complexity, "symbolism," and even "abstract art": the European Upper Paleolithic revolution rightly transplanted to southern Africa, so to speak.

But the problems that plagued the revolution in Europe were not resolved by its transfer to Africa (McBrearty 2007). The claims for the innovativeness of particular technologies were in some cases overdrawn. Stone blade technology is a case in point. Blades are carefully defined by archaeologists according to their morphology and production, and blade making was once thought to be *the* representative lithic technique of human modernity, the chief innovation of "Mode IV" technology, according to a typology from the 1960s, preceded by Levallois (Mode III), Acheulean (Mode II), and Oldowan (Mode I) technologies (for these, see chapter 5). But archaeologists have come to understand that blades were produced already in certain kinds of Levallois facture, and the research McBrearty and Brooks gathered showed that they reach back in Africa to a period long before *Homo sapiens* existed (see also Johnson and McBrearty 2010; Gamble 2007). Another supposed criterion of human modernity that has faded in recent years is pigment processing from ocher and other materials. The paintpots and incised ocher of Blombos are eloquent artifacts, but the scatter of ocher fragments at numerous older sites suggests that this material was collected and used for some kind of coloration (as well as for other purposes, perhaps) not only by presapients in Africa but also by Neandertals in Eurasia (McBrearty and Brooks 2000; Barham 2007; Roebroeks et al. 2012).

Moreover, much as it had been in the European scenario, the sense of postrevolutionary momentum and progress was exaggerated in the African revolution. As meticulous dating of the relevant sites has shown, the Still Bay and Howiesons Poort industries did not represent the march of modernity in southern Africa but were separate developments. Each seems to have flourished for fewer than 5,000 years, to be cut off by unknown forces, perhaps involving climatic and ecological change; and the second arose only after a hiatus of about 7,000 years (Jacobs et al. 2008). It is always good, at distances of dozens of millennia, to stop to ponder what seem like short intervals within them. These industries had their heydays at a chronological distance from one another that is two-thirds as long as the entire Holocene—or, put another way, more than twice the span that separates us from the Trojan War. The population that produced the second was likely different from the one that had produced the first, after millennia of migrations and resettlements; if the same general population was involved, it probably suffered some kind of demographic collapse or was forced to abandon the area for thousands of years in the interim. All in all, the situation seems to have been one of discontinuity and distinct populations, somewhat like that proposed for the European Aurignacian/Gravettian succession.

Although there were no Neandertals to underestimate in positing an African revolution, it was easy to do the same with early *Homo sapiens*, modern humans from the period before the Still Bay industry. Transferring the revolution to Africa might push back the clock for modernity by 30,000 years or more, but it still depends on a posited leap in capacities beyond those of preceding humans. This was exactly the kind of thinking McBrearty and Brooks had opposed in their marshaling of African evidence. The point of their accretive gradualism was to contest not a specifically European revolution but *all* sudden transformations in the behavioral development of Pleistocene humans, and their strong case for the variable profile of human behaviors for over 100 millennia before the 100,000-year mark weakens the idea of a unified leap forward, emphasizing again the complex, kaleidoscopic developments of numberless populations ranging through the huge continent where all hominins had gotten their start.

Finally, if a behavioral leap in Africa ca. 70,000 years ago *did* happen, what caused it? Those who champion the idea often have recourse to invisible biological developments, genetic alterations that suddenly gave our lineage the capacity for modern language, symbolic thought, and so forth, but that left no fossilized traces (Klein 2000). These are in general not compelling hypotheses, for reasons that I return to below but that can be previewed here in short form: complex behaviors of modern humans (language, symbolism, etc.) cannot be narrowly connected to determinism by a single gene or even a small set of genes (see also Gibson 2007; McBrearty 2007). Most archaeologists today attempt to resolve the puzzle of our modernity with thinking akin to that of McBrearty and Brooks, and the idea of revolutions in our prehistory finds less favor than it once did.

:3:

We can now put the puzzle of modernity in starker terms: anatomically modern humans appeared about 200,000 years ago, but there is little evidence for another 100,000 years of their acting much like modern humans. What explains this discrepancy between biology and behavior? It perplexed archaeologists as soon as the approximate birthdate in Africa of *Homo sapiens* was determined, and it was soon dubbed the "sapient paradox" (Renfrew 2008). To resolve the paradox, we must take seriously the varied, gradual accretion that McBrearty and Brooks proposed as a model for *Homo sapiens*' first 100,000 years or so. But we need other concepts as well.

In 2005 a conference in Cambridge, England, was devoted to "rethinking the human revolution." This was a follow-up to one held there eighteen years earlier, in which the European Upper Paleolithic revolution had assumed a place of honor. But the new gathering could more accurately have been qualified as "repudiating" or "rejecting" that revolution, and the counterrevolutionary consensus it represented is everywhere evident in the proceedings published two years later (Mellars et al. 2007). The prevailing sense gained from these reports is that, in the years around 2000, a new vocabulary had taken hold for describing the formation of human modernity. This echoes the terms of McBrearty and Brooks, and there is much agreement on the gradual, incremental appearance of modernity in Africa.

Discontinuities

But another aspect of the new vocabulary, somewhat at odds with the first, is an emphasis, greater than before, on local environments, discrete populations, and historical *discontinuity*. Our modernity was not merely a composite assembly, paleontologist Chris Stringer (2007) argues, but a development of independent innovations in behavior, some persisting and some disappearing without issue, in separate populations. Christopher Henshilwood (2007), the chief excavator of Blombos Cave, echoes this idea that innovations in the archaeological record might often have been dead ends. He highlights the stratigraphic (hence historical) hiatuses in the artifact record at his site and portrays modernity as a "mosaic"-like series of local events, with many "material culture cul-de-sacs"—though he also aligns this discontinuity with an idea, borrowed from Michael Tomasello, of cultural development as an accumulating bootstrapping, or "ratchet," effect. Lawrence Barham (2007), a student of the oldest African pigment use, tracks the "intermittent innovations" in that area over 200 millennia, while Francesco d'Errico and Marian Vanhaeren (2007), leading experts on Paleolithic bead making, frame its history as one of appearances, disappearances, and reappearances of techniques across a wide swath of Africa and Eurasia, dependent on historical contingencies affecting separate groups. A group of archaeologists working in Morocco sees an alternation of habitation and abandonment of sites there and correlates them with climate change (Barton et al. 2007). And archaeologist John Shea rejects proposals of cultural continuity in the Levant and ascribes the artifactual similarities that have encouraged them to "adaptive convergence among hominins occupying similar ecological niches" (2007, 228).

All this talk of discontinuity, moreover, was not limited to the Cambridge conference. Nicholas Conard, for example, the lead archaeologist for the German excavations that have yielded Aurignacian flutes and figurines, wrote in the same years a general account of our late-Middle and Upper Pleistocene prehistory that portrayed behavioral modernity as the coalescing practices of "regionally specialized social groups with highly variable material culture" (2007, 2031). In his view this piecemeal development, which he too likens to a "mosaic," occurred slowly from 80,000 to 40,000 years ago in Africa and Eurasia. Even as he painted a picture of transsapient coalescence, Conard thus worked to preserve the idea of cultural independence for independent groups, extending the technological localism and the variability of behavior that the Cambridge conferees also recognized. The need to do so must have seemed particularly pressing to a researcher devoted to novel European developments at the earliest moment of the Upper Paleolithic period—the *locus amoenus* for dreams of a European revolution, after all. Conard's position is that, while it remains true today that European sites have yielded kinds of artifacts that have no parallels dating back as far in Africa—musical instruments, animal and human figurines, and representational paintings—we do not need a Eurocentric revolution to explain the fact.

The general lesson to be drawn from this, and from all the recent positions highlighting the independence of ancient human populations and the discontinuities in their developments, is that the Afrocentric consensus does not nullify differences among early humans. We need explanations for these differences, and seeking them raises two consequential points about weighing the evidence. First, while it is always true in the study of the past that ostensible difference might only reflect fragmentary knowledge, *this does not rule out other possibilities*. Correcting for "preservational bias" might make much difference disappear, and archaeology can usually not rule out this null hypothesis. Perhaps Africans painted animals on rock surfaces before Europeans did and we don't know it, since they painted on outdoor sites rather than in caves, and their work has not survived; and perhaps African musical instruments and animal carvings dating from 60,000 years ago, long before the earliest European ones, will be unearthed next week or next month. Notwithstanding these possibilities, however, we must leave room, among our explanations for exceptional artifacts and behaviors, for a fundamental capacity shared by all modern humans: the capacity *to produce cultural difference* in varying social settings and in response to changing environmental affordance and demands. It might well be true that some European traditions 40,000 years ago were simply *that* different from African ones; but the flip side of this truth, then, would be that African traditions, the richness of which we are just now beginning to understand, were *that* different from European ones (McBrearty 2007).

Cultural difference, genetic difference

This brings us to the second, thornier point, which will require somewhat longer consideration. Signals in the archaeological record of human cultural difference have often been mistakenly thought to betoken differences in the fundamental capacities that enable culture in the first place, reflecting, then, genetically determined differences separating whole populations. The old narrative of the Upper Paleolithic revolution in Europe shows that archaeologists have made this mistake in the past, and their new analyses of discontinuity and variability represent in part their effort to construct hypotheses that avoid it. At root, the error is the same one sociobiologists and evolutionary psychologists fell into (see chapter 3): the view that only adaptation matters and that every phenotypic gesture of humans must somehow manifest our selected genotype in a straightforward correlation. In this gene-centered, adaptationist view, the presence of representational painting would reflect genetic determinants that are lacking where it is absent. But of course there is no gene for representational painting. It is the outcome of many processes involving not only multiple genes and histories of their selection but ontogeny, sociality, culture and its systems, negotiations with niche affordances, and other factors. Anything close to one-to-one correspondences between a gene (or even a few genes) and human behaviors of even modest complexity must be fanciful.

Genotype-to-phenotype mediation

The fundamental truth about complex animal behaviors is that they open a space between the genotype and its phenotypic expression, a place of mediation, negotiated with environment and circumstance, wherein genetic determination of response gives way to genetic determination of *the capacity for varied response*. In animal behaviors of modest complexity, this space might amount to little more than wiggle room: the possibility of two behavioral options in certain frequently encountered circumstances, for example. But the space widens according to the complexity of nervous systems and the behaviors they sponsor, and as it does so, what increases is not simply a tally of predetermined behavioral options but the possibilities for weighing, mixing, and mediating possible options. This enables a spectrum of responses to the environment, and so we have come back around to the complexity of animal niche construction and the feedback effects, over time, of behavior on the genome. Animals with highly developed nervous systems do not run their genetic identities at full throttle at all moments, as we might suppose of a paramecium or planarian. Instead, they respond through the mediation of an environmental stimulus that creates the interpretant and semiosis; and they deploy, in the mediated variability of this response, the capacity to mold action flexibly within certain limits or even to refrain from possible actions and thus hold behavior in reserve.

Animal cultures, then—the next step in the behavioral complexity I followed in chapter 4—have the effect of widening the opening between biological propensity and action in the world, and the limit case, the widest opening of all, comes in the most highly developed cultures, those of late hominins. What sets them apart from other animals, even other cultural animals, is the latitude granted by genetic predisposition to *counter* genetic predisposition; the store of behavior-in-reserve is vastly expanded.

And the most powerful ways of all in which hypertrophied culture in hominins came to widen the distance between genetic determinism and behavior were those described in chapter 5: its deepening archives and its formation of systems.

Cultural variability Cultural *variability*, in short, is what the evolution of our genome has determined in us, and given this history, we will always risk grievous error in assuming that genes could determine any cultural *fixity*. This is why attempts to tabulate general differences among present-day human populations in cultural complexity and the capacity to produce it, and then to correlate these differences with hypothetical evolved genetic differences, are wrong and pernicious (for recent examples, see Cochrane and Harpending 2009; Wade 2014). What makes them so is not their assumption that natural selection on humans continued long past the end of the Paleolithic era. This is demonstrated by well-documented instances such as selection for lactose tolerance in societies of milk-giving animal husbandry and malaria resistance in some tropical African societies. In these cases, one-to-one correspondences between a gene and a simple trait have been established and matched to societal histories that rendered the trait advantageous and caused the gene to be selected (Hollox et al. 2001; Itan et al. 2009; Tishkoff et al. 2001; for a more complicated example involving skin, eye, and hair color, see Wilde et al. 2014). Ongoing genetic evolution of *Homo sapiens* long after the Paleolithic era is now well established.

Projects aiming to discern population-wide differences in the capacity for and complexity of culture go astray on account of assumptions different from this. In the first place, they typically disregard the facts that genetic difference between humans is mostly a matter of individuals within populations, not of distinct populations, and that our species as a whole is, in comparison with our closest living relatives, genetically homogeneous to a high degree (Lewontin 1972; Lewontin, Rose, and Kamin 1984). The fundamental question of how we might define a discrete human population, in other words, is not readily answered by tracking genetic markers of a haplogroup. In the second place, such projects usually rely on specious sociological calibration or anthropological generalization in order to characterize the cultural differences they hope to explain. One of the leaders of this kind of thinking once confidently assured me that there is no complex music in a certain, large region of the world, when, from any considered cognitive and musical standpoint, the truth is rather that there is no human music anywhere that is *not* immensely complex (see Tomlinson 2015). Meanwhile, a recent book arguing for selected differences in intelligence among modern human populations relies on the naive (if not notorious) premise that

IQ tests measure intelligence reliably across all populations—that is, that they are not fundamentally biased by factors having nothing to do with the intelligence they purport to measure (Cochrane and Harpending 2009).

But most of all such projects are mistaken in their assumption that genes could in the first place direct culture this way or that, instead of providing only the biological foundation for an inevitable, flowering diversity. They ignore the mediated distance between the genotype and the complex behavior on which all human culture is founded, a distance that has been uniquely extended, beyond that of other animals, by our evolutionary heritage. In recent years, especially since the mapping of the human genome, geneticists have been particularly prone to this error, leaping confidently to adaptationist stories about the genetic determinism of modern human capacities (Nielsen 2009). An infamous example arose in 2005, when research indicating ongoing selection in humans of alleles of two genes indirectly associated with microcephaly led those involved to construct a tissue of unsupported speculations and to spin from them an evolutionary story with explosive implications. The scientists assumed that the alleles resulted in larger brain size and that larger brain size is correlated with greater intelligence, which led them to propose that selection on the alleles had been driven by the advantages of increased intelligence. The fact that the alleles are marked especially in non-African populations, then, encouraged them to historical speculation that their increase in frequency could be tied to two historical episodes: the advent of modern humans in Europe 40,000 years ago and the rise of domestication, cities, and writing in Eurasia from 10,000 to 5,000 years ago. Their conclusion was that they had offered genetic evidence for a selected increase in intelligence that could explain the supposed civilizational advantages and successes of the non-African populations in which they operated (Evans et al. 2005; Mekel-Bobrov et al. 2005). Here was a particularly virulent reassertion of a Eurocentric Upper Paleolithic revolution, with a Eurasian Holocene success story appended for good measure.

But the tissue of speculation could not hold. As Sarah Richardson (2011) has shown in analyzing the episode, there is no solid evidence that the alleles in question lead to bigger brains, and, more broadly still, there is no established correlation in modern humans between intelligence and brain size. Several research teams were quick to point out these shortcomings, including a team with some of the original researchers on it; the title of their article was unequivocal: "The Ongoing Adaptive Evolution of *ASPM* and Microcephalin Is Not Explained by Increased Intelligence" (Mekel-Bobrov et al. 2007; see also Woods et al. 2006). Richardson also rightly noted that the correlations the scientists drew between allele selection and historical events were too imprecise to be meaningful, but it is not only imprecision that troubled their history. The geneticists were dabbling in archaeology and getting it wrong: adducing a Eurocentric revolution at the very moment archaeological consensus repudiated it (McBrearty 2007), and localizing the Holocene advent of domestication and cities in southwest Asia just when an understanding was solidifying that these developments took place independently and almost simultaneously at several sites around the world (Price and Bar-Yosef 2011). The geneticists had been emboldened to construct their story, ultimately, because of their misplaced notion that genes not only can determine such things as lactose tolerance but also can micromanage the breadth and richness of culture.

: 4 :

The archaeological record of cultural variability reaching back to the beginning of our species shows our capacity to produce cultural difference to be a deep-seated one; this too is a consensus prehistorians have reached. But the account I have offered above signals a generative tension in the consensus. The McBrearty and Brooks model posits a cumulative assembly of modern behavior that began even before the full anatomical modernity of our species, while, at the same time, recent advances in fine-grained knowledge of the activities of early sapient populations paint a picture of separate cultural developments, intermittent innovation, cultural dead ends as well as thoroughfares, and convergences determined not by contact and transmission but by independent response to affordance and circumstance. The vocabulary to describe deep human history is divided between a cumulative, gradual achievement of modernity and a radical disjunction and discontinuity. The one seems not to mesh with the other, and together they plot no clear path from anatomically modern humans *not* acting like us to modern humans who do. So the sapient paradox remains unresolved.

Modernity repudiated? Perhaps, archaeologist John Shea has suggested in an attempt to cut the conceptual knot, this is because there never was any such thing as human behavioral modernity in the first place. Shea's manifesto of 2011, "*Homo sapiens* Is as *Homo sapiens* Was," strikes at the heart of the sapient paradox, pronouncing it to be a false problem driven by slow-to-perish archaeological ideologies. The idea of our modernity, for Shea, is a vestige of the old, Eurocentric archaeology that had sponsored notions of the Upper Paleolithic revolution. The historiography that lay behind this was the progressivist philosophy of irresistible teleology, liable, in mapping human differences, to conceptual abuse and racist reductivism. Both should be repudiated and the vocabulary of modernity itself abandoned. In its place we should follow a "uniformitarian" principle, according to which we presume that the earliest *Homo sapiens* bore all the universal capacities found in today's *Homo sapiens*. As evidence for this uniformity, Shea offers a census of lithic types (the four "modes" mentioned above) at late presapient and early sapient sites in Africa, which shows them to be widely and fairly evenly distributed.

Chief among the capacities ever present in our species is the ability to shape behavior flexibly in changing circumstances, and in emphasizing this variability Shea is a part of the archaeological consensus I have described. For him the capacity for variability, exercised "strategically" according to cost-benefit reckonings in the midst of particular environmental conditions, explains all differences in human behaviors. Where behaviors like those of more recent humans do not appear, either they weren't needed or their costs outweighed their benefits or, because of biased preservation, we haven't (yet) unearthed their remains. Shea extends this reasoning very far, to embrace even the ever-problematic category of symbolic behavior, which he supposes to have been fully developed in the earliest *sapiens*. In sum, today's variable behavior did not burst upon us in a revolutionary explosion, and it did not creep up on us as we gradually put things together. It was with us all along, a behavioral manifold in place from the beginning, part and parcel of our biological emergence 200,000 years ago.

Shea's argument seems a compelling one, and a distinguished group of archaeological colleagues responded appreciatively to his article, mostly using the (unresolved) dichotomous vocabulary of accretive gradualism and mosaic discontinuity discussed above. But I think the argument is mistaken on several grounds. First, Shea's reliance on preservation bias as a positive argument works a questionable twist on archaeology's null hypothesis: it no longer asserts that we cannot rule *out* behaviors because their remains may not have survived; instead, guided by the uniformitarian principle, we must rule *in* behaviors not found in the archaeological record. Next, the uniformitarian principle that founds Shea's reasoning, though it may be less prone to abuse than the progressivism of the older prehistorians he opposes, is just as much a leap of faith. By displacing the fact of sapient difference back to the 200,000-year mark, we do not escape the question of our difference and (by another name) modernity. When more fossil evidence is uncovered for the transition of *sapiens* out of our immediate forebears—the recently unearthed fossils at Jebel Irhoud, Morocco (Hublin et al. 2017) provide just this sort of evidence—will we extend the uniformitarian principle back 250,000 years? 300,000? Where does uniformity end? There is, in addition, also this evidentiary difficulty: for now, despite Shea's lithic analysis, there is not ample support for the degree of uniformity he extends back to sapient origins. To the contrary, as more and more early sapient sites are excavated, evidence mounts for the pattern described above of cumulative but discontinuous, mosaic-like change. It is true enough that those African flutes and figurines from 60,000 years ago that I imagined being unearthed in the near future might even be found in 200,000-year-old strata; but the likelihood diminishes with each site of that age that yields nothing of the sort. Never say "never," but sometimes say: "in all likelihood, not."

Ongoing gene-culture coevolution

The most general problem with Shea's argument, however, concerns evolutionary concepts. In rendering simultaneous the advents of our present-day biological and behavioral makeups, Shea ties the capacity for variable behavior narrowly to biology and genetics. There is an irony in this, since Shea's unexceptionable argument is for a great strategic variability in sapient cultures; yet he leaves little room for the yawning space of mediation I described between genotype and phenotype in animals with complex culture, and therefore little room for culture's feedback effects through niche construction on the genotype. "Behavior and its tangible by-products 'evolve' only in a metaphorical sense," Shea writes (2011, 2), discarding not only the particular mechanisms discerned by the niche constructionists but gene-culture coevolution and the whole thirty-year-old study of biocultural evolution. The possibility of a gradual change in the *scope* of sapient variability, induced by culture but finally, through the mechanisms of niche-constructive feedback, coming to be genetically based, is the first casualty of his uniformitarianism. I have argued instead that this is exactly the possibility we must entertain in order to resolve the question of early sapient evolution.

In truth, natural selection acting on *Homo sapiens* across the last 200,000 years cannot be dismissed; I have, just above, adduced two very recent instances involving lactose tolerance and malaria resistance. We must agree with Shea that early *sapiens* were clever creatures, responding flexibly to the demands of their environments; and we can agree also that they were unique, or at least close to unique (remember the Neandertals!), in the extent of their abilities to do this; but we cannot agree, I think, that their evolution through natural selection had come to a standstill. In this selection, and exactly *because of* the capacity for variable behavior Shea celebrates, culture played an unprecedented and highly complex role. Sometimes it had a buffering effect of the sort I described in chapter 3, minimizing the pressures of selective advantage or disadvantage on a population; sometimes it accelerated the process; and often it redirected altogether the nature of selective advantage. What stands behind this complexity is another force that arose from the capacity Shea stresses: the formative impact of these variable, capable humans on their niches. To resolve the sapient paradox, we require a model that accommodates early sapient behavioral variability and strategic canniness in the face of the world—features that Shea rightly emphasizes—but that recognizes the cultural role in niche construction and allows for the ongoing effects of natural selection.

: 5 :

A model for sapient evolution

I will describe this model here generally and in chapter 7 in exemplifying detail. It assumes determinate starting and ending points. It begins with the emergence by about 200,000 years ago of a species that deployed in its niche construction a novel flexibility of behavior, and it ends with a single lineage of that species some 70,000–60,000 years ago: the founding lineage of modern humanity, minimally but effectively differentiated from those around it, endowed in all essentials with our behavioral panoply, and setting off on its global expansion. The end point is easier to gauge than the starting point, since it is genetically marked in all human haplogroups today, as outlined above. The evidence that archaeology supplies also affirms the similarity of these humans to us in the sophistication of their behaviors. Three rough-and-ready instances, from three continents and each dating to more than 40,000 years ago, may suffice as tokens of this: ostrich eggshells from Africa carved with designs like those of modern San water vessels (Texier et al. 2010), the seafaring technology that was necessary to colonize Australia (O'Connell and Allen 2007), and flutes and figurines from Aurignacian Europe (Conard, Malina, and Münzel 2009; Conard et al. 2015). Natural selection on our species did not cease with the formation of the founding lineage and the coalescence of its modernity, but it was slowed (Chiaroni, Underhill, and Cavalli-Sforza 2009; Coop et al. 2009), and for reasons I will come back to, its force to remake our fundamental capacities to produce cultural complexity was decisively curtailed then. From this point forward, but only from this point, we can agree with Shea that *sapiens* was essentially as *sapiens* is today.

The starting point for the model is less transparent. As most who look back on our differentiation from other hominins agree, it requires a speculative leap: the assumption that our emergence must have involved the appearance of novel cognitive capacities. These changes need not have been dramatic, and indeed, the archaeological record in Africa, with its gradual diversification of techniques and innovations reaching back much more than 200,000 years, indicates that they were not. This archaeological picture of incremental emergence, arising in patchwork fashion across the huge continent, is supported by paleontological evidence, in particular the 2017 dating of Moroccan fossils of *sapiens*-like (but not fully modern) hominins to over 300,000 years (Hublin et al. 2017). The emergence of our species evidently involved a long history reaching back well before the 200,000-year mark and involving populations active in many regions of Africa. Whatever the cognitive features were that marked *sapiens* around 200,000 years ago, we can visualize them as modest changes when judged against the backdrop of our immediate forebears and perhaps even of other, more distantly related African hominins: changes that emerged from the feedback cycles of, and were selected on the basis of their efficacy in, the niche construction that drove all hominin evolutionary change. In this view, the novelties of early sapient cognition amounted to tweaks in the hominin "computational phenotype," perhaps subtle alterations in kinds of memory and in the capacity for its storage, which rearranged the relations of cortex to deep-brain processing.

Effects of cumulative culture Modest as they must have been, the changes nevertheless set in motion cultural patterns that would ultimately bring sweeping consequences, and I suggested in chapter 5 three directions these took: a tendency to sediment cultural archives in ever-deeper layers—a universal feature of modern humans that many observers have noted (e.g., Henrich and McElreath 2003; Boyd and Richerson 2005; Mesoudi 2011; Sterelny 2012); an increased ordering of learned materials into nested hierarchies of memory and experience; and the formation of these into cultural systems that abstracted behavioral response from stimulus and face-to-face interaction. Accumulation, systematization, and abstraction of cultural archives: these were the hallmarks of the culture that sapient cognition would create, and together they worked to widen the distance between genotype and phenotype and thus expand behavioral flexibility and variability.

Of the three, it was the tendency to accumulate culture that was a foundational teleology in sapient biocultural evolution, a vector powerfully directing our recent evolutionary career. It determined that sapient humans would deepen their cultures in ways that outstripped by far other animal cultures today—which transmit behaviors at a consistent, shallow depth—and that outstripped also those of all our extinct relatives, including Neandertals. Positing this telos does not mire us in the progressivist metaphysics that Shea rightly decries, for it is exactly the opposite of metaphysics, an interplay immanent in material affordances and constraints. The telos was not a supramaterial aim but an attractor, an emergent tendency of a certain kind of computational machine—sapient cognition—at work in its taskscapes. The machine sedimented cultural materials because this was the comfort zone of its operation, the point to which it gravitated in the space of its possible effects. The sapient archaeological record, which presents in synoptic perspective a gradual accumulation and gathering of complexity, is the revelation of this attractor and of the machine at work. The consensus view of this record among archaeologists locates the start-up of the machine at least as far back as the beginning of our species.

The computational model is just that—a model—and it certainly does not answer all the questions we might ask about this novel sapient cognition. (What was the exact nature of the cognitive tweak? What conditions in African niches selected for it? Why didn't other hominins, especially Neandertals, come to be able to accumulate cultural materials to the same depth?) But once the pattern of cultural accumulation took hold in a situation of niche construction, it could never have been a neutral piling on of behaviors, social patterns, and ideas; instead, it always altered the niche in some manner and degree. Extended across dozens or hundreds of generations, the general effects of the accumulation were dramatic. In the first place, the complexity of cultural order and system tended to grow, all other things being equal, since this is how the cognitive machine shaped and managed the accumulating archive. With increasing elaboration of cultural systems came an enhanced impact of culture on the niche, and taskscapes of broadened scope took shape. As these taskscapes were more and more thoroughly transformed by culture, successful habitation of them grew to depend on the transmitted cultural archive, since its forms guided humans more and more effectively through their negotiation of environmental affordances and challenges. This brought selective pressures to bear, and those groups or members of groups most able to cognize the archive and exploit its systems gained advantage over those less so. Selection favored genetic determinants of the mechanism sponsoring accumulation in the first place, but it did so because cumulative culture had shaped a niche whose viability depended in part on—cumulative culture.

The advantages gained through cultural work determined local optima for early humans in their niches, to recall again the vocabulary of Van Valen's (1973) Red Queen model. Niche transformation through cultural means enabled them to perch on an adaptive peak in their coevolutionary competition for resources with the organisms around them. But the perch was always precarious and the peaks were always shifting, so behavioral variability was early humans' most effective armament. As the variability was expanded in the formation of deep, systematic cultural archives, local societies equipped with them surpassed in some measure those without them. Perhaps this was a modest measure, barely visible; I noted before that modeling of Neandertals and sapients in Europe points to large and rapid effects of even small advantages, and the same might hold true for different sapient groups. In any case, here again selective pressures must have worked to enhance the cognitive capacities that enabled cultural complexity.

In describing the cultural niche construction of these early sapients I have reintroduced the term *taskscape*, which I borrowed from anthropologist Tim Ingold (1993, 2000) in chapter 5 to name the meeting space of signs, ideas, forms, systems, and behaviors, on the one hand, and the stuff of the environment, on the other. The long-term development of sapient culture across our first 150 millennia may be conceived in terms of the increasing power of human groups to build culturally fashioned taskscapes from more neutral ecologies and more neutrally inhabited landscapes. The niche-constructive model begins to indicate why it is implausible to think that the transformative scope of early humans working their taskscapes remained a static, unchanging force across the millennia. Instead, the feedback from cultural niche construction must have generated a selection for the enhancement of the capacities supporting this scope; all other things being equal, again, the range, power, and systematicity of human action must have tended to increase.

Discontinuities, again

But of course all other things were never equal. The picture of a smooth accumulation of cultural complexity brought about by sapient cognition-in-the-niche presents only a statistical generalization for the whole species, and we must not imagine it as a groundswell gently carrying forward each and every human group. As *Homo sapiens* followed its African course, it must have evolved in a dizzying array of fissioning groups, separate histories, migratory pirouettes, fraught encounters, interbreeding mergers, and new divisions. Along the way some local populations died off while others expanded their reach and divided anew; some wandered far afield while others moved slowly along with shifting resources and still others oscillated back and forth with seasonal or larger changes; always the panorama of the species as a whole showed ceaseless movement. This was enabled by the very capacity for behavioral variability that set the species apart, which facilitated its habitation of a wide range of ecologies—the conversion into taskscapes of many different landscapes, in other words. And the movement was impelled by the shifting ecologies themselves, altered slowly or abruptly according to the vagaries of late-Middle and Upper Pleistocene climates.

Archaeologists today understand that the discontinuities of behavior they reconstruct are the record of this incessant movement. In the growing number of case studies where it is clear that discontinuity is not merely an artifact of our fragmentary knowledge, these breaks are seen to reflect one of several demographic possibilities: the dying off or severe decline of a population, its movement along a gradient of shifting resources, or its migration away from a region turned inhospitable. The breaks must manifest, in one scenario or another, the inadequacy of sapient variability to rise to the local challenges at hand and climb a fitness peak; in these cases the Red Queen could not keep up. What is *not* conceivable in these cases, however, is that sapient humans ceased to be humans: that they failed to deploy, even in an effort finally defeated by larger conditions, the flexible and varied behavior that was their hallmark and characteristic response to the world.

: 6 :

We have come back around to the dichotomy of short-term, local discontinuity and *longue durée* accumulation. How does the one add up to the other? If the local picture and the global panorama are not merely disconnected views from two different perspectives but must be related in a biocultural evolutionary history, what mechanisms make the connection?

Cultural convergence and attendancy

Evident in the sapient record, as some archaeologists have noted, is a tendency toward the parallel development or convergence of independently shaped behaviors. Shea has proposed such parallelism to explain separate but similar habitations of the Levant (Shea 2007), and Marian Vanhaeren and Francesco d'Errico see bead making as the exercise of similar behaviors independently repeated across huge swaths of time and distance (Vanhaeren and d'Errico 2006; d'Errico and Vanhaeren 2007). I suggested in chapter 5 that bead-making systems would have hovered in *attendancy* over certain often-repeated intersections of technical proficiency, social complexity, and material affordance, and I offered bird-bone flutes as a second instance. In each case the system and its components created an attractor toward which human groups moved independently of one another, so that generally similar outcomes were repeated in separate circumstances. There is, once again, no magic involved in this attendancy, and no metaphysics; it springs from the immanent relations of materials, behaviors, and ideation—with, of course, sapient cognition connecting the dots. Given the coalescent force of the attractors involved, such convergence or behavioral parallelism must have happened often.

Across the long, mobile history of sapient humans, when two of these separate populations met, even after millennia of separation, the parallel paths they had taken could facilitate a confluence of cultural system and practice. The beads of the one group and those of the other—in general, the tokens of social difference both had come to fashion—opened a space of cultural recognition, a common ground that rendered encounters with other human groups, however far-flung, *familiar*. (An intriguing possibility pondered by some archaeologists is that a certain *lack* of familiarity marked instead the meetings of sapients with other kinds of humans.) Parallel cultural developments of separate sapient groups reflected similar kinds of cultural niche construction, and so the taskscapes that came together in encounters must have borne family resemblances to one another. And, as the example of beads reflecting social difference suggests, the familiarity was not limited to material technologies that have left their traces for archaeologists to unearth. Immaterial patterns of social organization probably also developed along parallel tracks, even while their details were defined by local context. Such phenomena as arrangements of rank and status, the growing intricacy of kinship constellations, and the emergent differentiation of ritual activity from normal activity, whatever their local differences, must also have facilitated recognition. And, as the advantages of precise, effective communication must have been pressing on all sapient taskscapes, in this expansive arena too there would have been a general developmental parallelism.

From cultural to genetic parallels It was not only culture that tended toward parallel development, however, for as it did so, it created conditions fostering parallel changes in the genotype as well. I am not suggesting, needless to say, that the general similarities of culture, sociality, and behavior could have brought about identical, specific mutations in the genotypes of the separate populations practicing them. But if we look back at the diagram of cultural niche construction at the end of chapter 3 (p. 57), we can discern outlines of a broader genetic impact that the cultural similarities would have had. Similar cultural patterns would have shaped taskscapes in similar ways, and these would have come to depend on the cultural behaviors that rendered them viable. Selection in a particular population, then, favored those who could best manage the more and more thoroughly encultured niche, that is, those who could manipulate the sediments and systems of culture to make the taskscape a successful one in exploiting available resources. Incremental genetic enhancements to these capabilities would have been favored in any population where the accumulating complexity of culture existed—which is to say, in all sapient populations.

This suggests that, once the complex divisions, migrations, and isolation of these populations had begun, the cumulative culture that marks *Homo sapiens* in unique degree would have set them all on the same general course, not only toward broad *cultural* parallelisms but toward a parallel *genetic* enhancement of the foundational capacities to make complex culture. We can begin to elaborate the example, raised above, of effective communication. Wherever sapient populations moved, greater communicative effectiveness would have increased the power of their taskscapes to exploit resources; so all these niches would have driven selection for genetic changes enhancing communication. This is the closing of the feedback loop in the diagram, whereby the genome altered by cultural niche construction alters in turn the capacities for that cultural work. It means not only that the parallel cultural developments in separate human populations inhabiting separate niches would have fostered convergent tendencies in the selection shaping their genomes, but also that these genomic changes would ultimately have redoubled the general parallelism in cultural depth.

The convergent development of separate populations of *sapiens* thus affected all three lines of inheritance that drove their evolution: cultural, environmental, and genetic. Niche constructionists, we remember, distinguished these three in their model and understood both their independence from and impact on one another. Lineages of cultural traits, of features of the constructed niche or taskscape, and of genetic variants did not cross over with one another—there is no Lamarckism here, and no human individual developing strong hands through flint knapping passed along stronger hands to her children—but the lineages moved forward in tandem, linked in feedback cycles. As they did so, they created parallel lines of cultural and genetic development in all sapient populations. These lines enhanced the basic capacities for cultural accumulation, elaboration, and variability that had been set in motion already 200,000 years ago, altering the scope of human action in the world across the following 150 millennia. At the same time, they facilitated the meetings and mergings of sapient populations that must have occurred repeatedly across this period.

Selection at a standstill

The niche constructionists' models indicate a number of specific effects that culture could exercise, through the mediating environment, on the changing genome. These included the acceleration of the spread and even the fixing of certain genetic variants in a population at the expense of others; the creation of switch-like thresholds marking points of behavior or resource availability, on either side of which a different allele might be favored; and the buffering of the population against selective pressures that might be felt in the absence of cultural adjustments, which in extreme form could even result in a counteracting of selective pressure, or "counterselection." The first and last of these can be thought of as opposed tendencies, in which changes in the environment brought about by cultural action either bolster or weaken a prevailing selective gradient; the possibility of threshold switches marks instead a middle ground, in which more complicated dynamics take effect. All three of these general dynamics were important in the history of advancing hominin culture, but the buffering effect played a special role in bringing large-scale genetic shifts in sapient capacities effectively to a halt around the time of the formation of our founding lineage, by about 60,000 years ago. To see how it did so, we need to look at it more closely.

This buffering is usefully conceived as a "relaxing" of the selective gradient, in which the selection of an advantageous trait is weakened because the advantage it confers is supplied instead by habitual behavior (Deacon 2010). The behavior in question need not be cultural, and a famous instance involving noncultural behavior is the inability of humans and some other primates to manufacture vitamin C, necessary to our metabolism. The loss of this ability is attributed to our arboreal ancestors' habitual fruit consumption, which provided an external supply of vitamin C. Since internal vitamin production was no longer required for survival, the selective constraints on genes controlling it were weakened, and, with the accruing of mutations that were no longer weeded out, genetic control and vitamin production both gradually degraded. If noncultural behaviors could thus bring about such relaxed selection, so also could the long-term, iterated patterns of culture, and so, all the more, could cultural patterns whose persistence across generations was stabilized and extended by their systematic arrangement.

The problem of the stable achievement of human modernity about 60,000 years ago can be considered a similar buffering—in this case, a cultural buffering—against genetic selection, a possibility that has been noted by evolutionists before (Chiaroni, Underhill, and Cavalli-Sforza 2009). But here it occurred with a twist, for this buffering effect was reflexive: cultural capacities and behaviors lessened the selective gradient by which their own scope had been enhanced over the previous 150 millennia. That is, *cultural capacity expanded to a point beyond which it buffered our species against selection for further genetic enhancement of cultural capacity.*

How could this happen? To answer the question we must remember, first, that our capacities for cultural accumulation and systematization do not constitute a "trait" in anything like the Mendelian sense of the word; they are not the product of a single gene or allele but result from a network of genetic controls and the developmental patterns and potentials that these set in motion. To think of them in Mendelian terms—like colors of pea flowers, vitamin C production, or, though they are genetically more complex, colors of skin—is the mistake made by those, discussed above, who imagine that the most general and foundational differences in capacities could exist between today's human populations. They take the selective processes evident in the relatively recent proliferation of such monogenic phenotypes as sickle-cell malaria resistance or lactose tolerance not merely as evidence of ongoing natural selection—in this they are right—but as evidence-by-analogy for ongoing selection that might alter fundamentally and sweepingly general human capacities—in this they are wrong.

To recognize the multiplex, mediated genetic determination of human capacities for cultural elaboration is to take a first step toward explaining why the buffering effect brought their evolution to a standstill. The second and crucial step depends instead on *what was selected for* in the long period leading up to the stable coalescence of our modernity. The watchword here is, again, Shea's variability. All observers agree that this is what sets modern humans apart from other living species and from our extinct relatives (with the usual, cautious qualifications for Neandertals); this is the broadest consensus in archaeology, paleoanthropology, ethology, and related disciplines. I have argued that, from the beginning of our species down to our founding lineage, cumulative culture, its systems, and its niche-constructive effects altered this variability, bringing about the selection of a genome that enhanced its scope. The effect of this enhancement of human variability was to widen more and more the distance between phenotype and genotype—the gap in which variability itself found its expression on the taskscape.

A tipping point or threshold was here approached, for, once behavioral variability had reached a scope that enabled humans to range across every continent and flourish in all environments except the most forbidding, variability itself could no longer engage with the Darwinian algorithm as an advantage to be selected. (There was, on the other hand, still ample room for behavior to alter the selective gradient and the genes controlling simpler, Mendelian traits such as lactose tolerance and malaria resistance.) Wherever humans went, they buffered themselves against selection for any greater cultural ingenuity by the operation of that ingenuity. On every taskscape and in every population, our adaptability of behavior—which is to say, our hypertrophied culture—was such that selective constraints never came to bear. At this point, the feedback impact of the genome on the general determinants of our cultural flexibility and variability was diminished to insignificance, and the phenotype/genotype gap could not be widened any further by the mechanisms that had opened it in the first place. A program of behavior of immense flexibility and complexity was fixed throughout the species.

If it is hard to think of another instance of natural selection bringing about such an outcome, that is because there probably are no other instances of selection creating so varying, change-producing, and self-adapting a phenomenon as human cognition and behavior had become. Cultural difference had placed itself beyond the reach of selected genetic difference.

: 7 :

Cultural systems and epicycles These, in general terms, are the features of the model of sapient evolution here proposed: its starting point, at the advent of *Homo sapiens*, in a modestly altered type of hominin cognition; the effects of the accumulation of cultural archives, unique in degree, that the new cognitive type enabled; the parallelisms and convergences, cultural and genetic, that arose in the niche-constructive workings of human culture, which reconcile archaeologists' panorama of steady accumulation with their detailed picture of discontinuity and disjunction; and the shutting off of genetic selection for increasing variability in our species, when cultural capacities themselves moved beyond the reach of selective mechanisms. And there is one more, special feature of the model, alluded to frequently above, that requires its own, concluding word: the impact of the systematization of human culture.

The discussion in chapter 5 of systems in late hominin culture laid the groundwork we need to understand their general roles in sapient evolution. In the broad view, the systematic nature of hominin cultures had always increased along with their deepening sedimentation, as was already evident, for example, in the difference between Acheulean and Levallois operational sequences. This tendency only strengthened with the deeper accumulations that *Homo sapiens* brought about. For all late hominins, advancing systematization of culture led to the progressive widening of the distance between phenotype and genome in hominin evolution; in *sapiens* this too was carried to an unprecedented degree.

The abstraction of sapient cultural patterns that resulted depended upon the emergent features of the systems. In the first place, their integrated structures, joining components from semiotic, behavioral, and material realms in hierarchic arrangements, set them off from less systematic cultural forms and gave the components new, emergent functions. This integration also fostered the stable, holistic transmission of systems in cultural practice, endowing them with a kind of system-wide functionality in addition to the component-by-component one: *metafunctions* too emerged with the systems' formation and persistence. These holistic metafunctions introduced in animal cultures a new level of complexity, creating lineages of evolving systems. The lineages of replicated and varied cultural systems created search spaces, cultural *morphospaces* with their own tendencies and attractors.

The forces loosed by this systematization pushed toward autonomy of the evolving systems from the conditions that gave rise to them. One expression of this is the generalization evident in the attendancy of certain systems on certain congeries of material, behavioral, social, and semiotic phenomena: bead-making systems were bound to take hold in the presence of *these* ingredients, flute-making ones in the presence of *those*, but the ingredients themselves were of the most general sort. The systems emerged from the ingredients but also redirected and transformed them, assuming an element of control over the very phenomena that spawned them. This controlling action of integrated cultural systems was redoubled in effect as they were passed along in evolving lineages. It enabled the systems to gain some distance from the feedback cycles of cultural niche construction and thus to function in the unfolding history of human culture as feedforward mechanisms. For this reason I have called such systems, the culminating manifestations of cultural autonomy, epicycles.

The coalescence of systemic evolution, autonomy, feedforward control, and epicyclic operation marks the apogee of sapient culture and, by extension, of all hominin culture, indeed of all animal cultures. Although cultural systems antedate *Homo sapiens*, as we saw in chapter 5, archaeological evidence does not suggest that all these additional systemic elaborations had fallen into place 200,000 years ago. A steadily growing body of evidence reveals their incipient formation about 100,000 years later, however, and abundant signals from disparate parts of the world show their secure operation after another 50 millennia, in the period of the dispersal of our founding lineage. These immensely consequential workings of cultural systems, in other words, define the formation and stabilization of our modernity. In chapter 7 I will speculate how they came to do so in five central areas: ritual, music, language, advanced technology, and metaphysics.

CHAPTER 7

THE GATHERING OF MODERNITY

:1:

Beyond language and symbol Two related ideas about modern humans have until now guided thought about our deep history. The first places language at the heart of our emergence, where it is lodged as the sovereign determinant of our natures and as a cognitive puzzle that, once solved, would illuminate the origin of our modernity all told. This emphasis on language is habitual in literature from several fields, including archaeology, cognitive studies, and linguistics; we can call it *linguocentrism*. The second idea adjusts the first by substituting symbolism for language as the universal human attribute. Advocates of this *symbolocentrism* define their key term with varying degrees of care. In the best accounts, symbolism is analyzed as a semiotic mode basic to language, while in less thoughtful ones, many sorts of artifacts or inferred behaviors of ancient humans are judged to be "symbolic" without much clarification. All, however, agree in proposing symbolism and symbolic cognition as the foundation from which arose the social and cultural complexity of modern humans.

These two lines of thought hold that we are not merely *Homo sapiens* but *Homo symbolicus* or *Homo loquens*. Explain the appearance of the one or the other, and you will explain how we became what we are. If I begin by asserting that either approach is a misstep, this is not to deny the importance of symbolism or modern language, which both must be central elements in any description of modern humanity. The archaeological record from 70,000 to 40,000 years ago suggests that they were present, essentially in their modern forms, by the time the founding lineage of today's humans formed in Africa, and the advantages they conferred, most archaeologists would agree, are reflected in the rapid expansion of this lineage across the world. But locating symbolism and language at the end point of our development, at a time when our modernity and general cultural capacities were fixed, is a far cry from taking them to be the driving force within it, as linguo- and symbolocentric accounts do. The course I have taken of scrutinizing hominin culture prospectively, from the bottom up, resists the temptation to define modern phenomena in this fashion as determining factors or to use them as contemporary touchstones from which to reverse engineer our history. Instead, I have sought to show how the phenomenon that is modern humanity, including language and symbolism, emerged amid the infinite array of possible courses evolution could have taken amid a set of conditions as complex as animal communication, sociality, and culture. We need only glance at the variety of animal communication in the world today, from hour-long gibbon duets (Geissmann 2000) to birdsongs, from honeybee dances to the individual- and pod-specific click-codes by which sperm whales identify each another (Gero, Whitehead, and Rendell 2016), to know that innumerable forms have resulted from as many selective histories. Insofar as symbolism and human language command our attention, our task is to understand how they became two of *our* distinctive forms in the gathering of capacities that constituted our modernity.

Countering unilateral narratives

The notion of a *gathering* of modernity spotlights from a different vantage point the shortcoming of linguocentrism and symbolocentrism. Both tend toward the kind of monocausal explanations that are ineffective as tools for building a deep-historical account of networks of feedback interactions. To start with the assumption that our evolutionary heritage can be usefully modeled as the formation of a single present-day capacity is to set off in the wrong direction; our a prioris must instead point toward a multiplex, nonlinear causality. The communicative modes of whales, birds, bees, and orangutans form parts of highly developed behavioral assemblages, but no single one of them could reasonably be thought of as predominant in the evolution of its assemblage, which instead took shape from the networked interplay of many capacities, phenotypic features, and niches. Similarly, however important human language and symbolism would eventually become, each must have emerged as one element in the network of our own behavioral assemblage, and in looking back through our deep history we quickly reach a time when they become indiscernible as lineaments of humanity. How quickly? The findings of archaeologists suggest that this time lies *within* the span of *Homo sapiens*, for there is little evidence weighing for, and ample evidence weighing against, the idea that the earliest *sapiens* possessed either fully developed symbolism or modern language. These late hominins likely constructed their taskscapes in more complex fashion than their predecessors, but they were not creatures already possessing language or symbolism or mysteriously destined for them. These capacities emerged in the course of the 150,000 years after the advent of anatomically modern *sapiens*, as parts of the gathering formation we must attempt to model.

In the preceding chapters I have put together a conceptual toolkit to build this model. From evolutionary theory come the possibilities that arise with cultural niche construction, which I outlined in chapters 2 and 3 and applied to the early sapient situation in chapter 6. Here the guiding propositions were that the effects of culture can alter our species' selective terrain, and that from this follow all the particular intricacies of buffering, acceleration, thresholds, and equilibration that manifest those effects. Chapters 5 and 6 described a set of additional dynamics, which arose from the complex systems of late hominin culture as outgrowths of the accumulating archive: the attendancy of certain systems on recurrent constellations of semiotic, social, and material conditions; the resulting pull toward parallel or convergent developments in sapient groups, however independent; the holistic transmission and abstraction of cultural systems themselves, and through these processes the emergence of system-wide autonomy and functionality; and epicyclic, feedforward effects that arise in the most autonomous cultural systems.

Finally, from the broad horizon of animal semiosis comes a set of conditions involving semiotic process, sign types, and the organizing of signs into systems that I laid out in chapter 4. These conditions must loom large in any deep human history, not merely because semiosis forms the stuff that is learned and transmitted in all culture, but because the trajectory of early *Homo sapiens* cast such transmission in increasingly systematic forms. These did not at first involve symbols; they could not have done so, since an elaborated systematicity was a prerequisite for symbols to appear. Instead, systems took hold among signs that did not require systematic organization for their operation, chiefly indexes, and slowly yoked these into organized arrays. What I have called *hyperindexical* systems, as well as full-fledged symbolic ones, are universal, constitutive traits of *Homo sapiens* today, but they were not always so.

The deepening systematicity that finally gave rise to symbolism reflects the hierarchic nature of semiosis itself. Complex kinds of signs depend for their function on simpler signs, so that indexical animals produce icons also, and symbolic animals produce both icons and indexes. From this hierarchy arises the expanded semiotic flexibility of animals that command indexes or symbols. This versatile cognizing of the world lays the foundation, then, for other forms of behavioral, cultural, and social variability. And an unprecedented variability in all these regards, as we saw in chapter 6, came to be the hallmark of the hominin lineage late in its history.

: 2 :

The Indexical Age

Early sapient humans were adroit semioticians, deploying indexes to manage social lives at least as complex as those of the other hominin species around them, and more complex than those of nonhominin lineages. In this they looked back on a history that was already long, for indexicality came early to be the primary kind of semiosis in the social interactions of hominin groups. We can infer this, against the backdrop of the general indexicality of mammals and especially monkeys and apes, from archaeological evidence. Lithic industries and their transmission show that the learning skills and situational memories of hominins were already developed beyond those of other animals more than 1.5 million years ago. These capacities are prerequisites for constructing indexes, which differ from icons exactly in their reliance on learned and remembered associations involved in prior situations. Elaboration of these capacities must have conduced to the hominin elaboration of a more general animal indexicality.

This principle of elaboration encourages a sketch, in very broad outline, of the semiotic evolution of the hominin clade. Starting with the earliest hominins about three million years ago and positing for them a limited indexical sociality, perhaps not much different from that of a chimpanzee troop today, we can think of subsequent hominin evolution as the gradual flowering of an Indexical Age: a slow, sporadic growth across several or many species in the variety, intricacy, and efficacy of index use, which eventually outstripped that of all other animals. The growth was above all due to a set of feedback loops among culture, niche, and genotype. As hominins' cognitive capacity for shared attention increased, so did the coordination of social interaction, and the rough-and-ready turn-taking of modern human discourse began to form (Tomasello 2008). The kinds of innate indexical gestures and calls that had long characterized hominin and prehominin interactions (Burling 1993, 2007) and are still observed in primates today (Seyfarth et al. 1980) came to be linked to this coordination (Bowie 2008). The linkage of vocalization to coordinated interactions enriched the makeup of the taskscape, and the advantages of this heightened management of the niche drove selection for greater control and variety of vocalization. At first this could have taken the form of a buffering such as that described in chapter 6, in which genetic constraints on innate calls were weakened as culture offered alternatives and so altered the selective gradient; the calls were, in part, "offloaded" to culture (Deacon 2010, 2012a). New capacities for producing more varied and controlled vocalizations could not have come online, however, without new modes of perceiving and interpreting them, and selection for the calls enhanced also the shared attention that had set off the communicative enrichment in the first place. A feedback circuit was closed, and indexicality was heightened in tandem with the growing intricacy of sociality.

Early indexical culture In this way, long before symbols existed, indexes came to be a rich, malleable resource for managing the constructed niches of hominin life. Their pointing or indicating capacities and clear causal implications were well suited to situations of intimate copresence on the taskscape, since they conveyed transparent meanings: actions to be undertaken (or not), moves to be made (or avoided), and affective responses to unfolding situations. Ethological observation of nonhuman primates today indicates this, but it is also once again an inescapable inference from archaeological evidence. In particular, a group of well-studied European sites dating back 500,000 years or so and associated with *Homo heidelbergensis*, the common ancestor of *sapiens* and Neandertals, testifies to the precocious development of indexical culture. These sites, which include Boxgrove in southern England, Bilzingsleben in eastern Germany, Aridos in central Spain, and Notarchirico in southern Italy, have inspired disparate interpretations. Were the carcasses butchered at Boxgrove and Aridos hunted by the hominins involved, which would indicate that these groups were capable of bringing down large prey, or were they merely scavenged? (The consensus today is: hunted, at least sometimes.) Were the hominins at Bilzingsleben able to build structures or not? (Perhaps; the evidence is difficult to read.) Were activities organized around hearths or decentered and more haphazard? (Sometimes around hearths.) All the interpretations, however, even the most cautious, point to social and communicative interaction and taskscape building of considerable sophistication.

The Boxgrove excavations, for example, cover a large area, comprising sites visited repeatedly by hominins as well as sites of one-off butcherings of horse, rhinoceros, hippopotamus, and other animals. Even in the most ephemeral, opportunistic meetings, archaeologists discern here intently structured sociality and exploitation of resources (Pope and Roberts 2005; Gamble 1999, 2007). For the butchering of one horse, eight Acheulean bifaces were fashioned quickly from cores gathered from a nearby cliff. The cores were tested at the bluff for their suitability before being carried to the carcass; others were tested but deemed unsuitable and discarded there. Larger stones and bones already cut clean were pressed into service as anvils for crushing and surfaces for cutting, and much of the harvest must have been consumed on the spot, probably raw, as there is no evidence of fires for cooking. The interaction would have been quick, intense, and fraught with danger, since four-legged scavengers and predators would also have been attracted to the kill. Decisions had to be made, negotiating "tensions between the functional need for butchery tools, effort in transporting unmodified raw material to a kill site, risks to the individual in leaving a large group, and keeping a carcass secure" (Pope and Roberts 2005, 85). The bifaces were carried away when the butchering was done, but at other sites, probably places of ambush of prey that the hominins repeatedly exploited, deposits of them accumulated; these would have provided a ready-made supply for future butchering.

For Boxgrove excavators Matt Pope and Mark Roberts (2005), the evidence points to a landscape structured according to routinized, cultural uses and even marked durably and recognizably by these routines—a highly elaborated taskscape, in other words; and they propose that the beginnings of such usage can be traced back a million years farther at certain African sites. By the Boxgrove period, the interactions on such taskscapes were intricate "performances" (Gamble 2012) in which many social dimensions intersected in even the most urgent situations: distribution or hoarding of foodstuffs, on-the-fly negotiations of these acts involving strong affective response and assertions of status or submission, and a situational variety that spawned transient divisions of labor, as some individuals collected and tested stones while others fashioned tools, sliced meat, crushed bone, or stood guard. Half a million years ago such complex performances unfolded at locales scattered through Africa and Eurasia.

In these kinds of hominin interactions, the index extended the reach of its indicative, deictic meanings. It is implausible to think that anything like modern language was within the reach of *Homo heidelbergensis* or any other hominins of the Boxgrove period or for a long time after it; so the first lesson we learn from such sites is that ancient prehumans could collaborate on complex activities of many sorts without the aid of modern language. The communicative medium deployed in its stead is usually termed *protolanguage*, and its putative features have been the subject of frequent analysis and speculation (for a review, see Tomlinson 2015). In such discussions the reverse engineering of linguocentrism often makes an appearance: the term itself suggests a medium *on its way to* full-fledged language, and the analyses, almost without exception, concentrate on language's modern features, especially the formation of a set of phonemes, a lexicon, and a syntax. In avoiding this route I use the alternative and more neutral term *protodiscourse*, coined by linguist Jill Bowie (2008) to connote a "sequenced communicative behavior" transmitting emotions and intent or indicating close-by things and situations. Whatever language-anticipating elements might have formed in protodiscourse, its foundational resource must have been a highly developed indexicality employing vocal and bodily gestures of considerable variety and eloquence.

Hyperindexical culture The hundreds of millennia that separate Boxgrove from the flourishing of Neandertals and the rise of *Homo sapiens* no doubt saw additional gains in the expressive range and power of such indexicality. The growth was repeatedly triggered, on taskscape after taskscape, by the selective advantages that accrued to efficacious communication and to the sophisticated sociality it made possible, which set in motion countless replays of the feedback mechanism described above. At the end of these selective histories, 200,000 years ago, the vocalized indexical repertories of Neandertals and early sapients probably employed more varied timbral possibilities, articulations, intonational shapes, and rhythmic patterns than before. All these are elements of what linguists call the *prosody* of modern speech, but we need to imagine them deployed long before words and syntax existed: as a culmination of the Indexical Age, not a harbinger of language. These indexes were by now less genetically controlled and more subject to cultural learning than they had been—a result of the buffering and offloading noted above—but this did not loose them entirely from genetic evolution; instead, their roles in culture were exerting more and more influence on the novel kinds of niche construction hominins achieved, with all its genetic repercussions. The indexes at this point may have involved some incipient conventionalization of the reference of particular vocal signs, but there is no compelling evidence to indicate that they were symbolic in the semiotic sense of that word or that symbolic cognition had come into its own.

This last point disputes the suppositions and hypotheses of many archaeologists, who have been quick to find symbolism in the traces they have unearthed. But symbols, we remember, require both an agglomeration or array of differentiated signs and a systematization that fixes relations among them. The fact of an arbitrary relation between a single sign and its meaning, while typical of symbols, does not create them. Instead, symbols arise when the internal organization of a sign system creates relations of indexicality among the signs in it and a metafunction for the system as a whole, which comes to be an index of collections of objects and actions outside it. I will return below to the precise nature of this external reference; for now it is enough to recall that it depends in part upon the structured interrelations within the system, and that this is a matter of a particular action of the interpretant, not merely of a particular connection between sign and object. The question of the presence or absence of symbolism among Neandertals or early *sapiens* does not turn, then, on identifying a specific behavior, such as body painting with ocher, or on finding a particular artifact, such as a piece of incised shell or a bead for hanging on a necklace. It turns instead on the interpretive complexity of the semiotic system that we can convincingly propose these behaviors and artifacts to have been part of.

Framing such proposals is the more difficult because the complexity of cultural systems forms a graded spectrum, not an either-or proposition, and so there is no clear threshold between symbolism and its absence. As I said in chapter 5, we might deem Oldowan technologies minimally systematic, we can think of Acheulean ones as modestly so, and we must think of Levallois techniques as more elaborately so—but in hominin cultures we cannot easily draw a clear boundary between system and nonsystem. Symbolism, a product of cultural systems, arose as an outgrowth of indexicality developed to a particular, high degree of systematicity. Across the Indexical Age, increasing variety and differentiation of indexes led, hand in glove with growing cognitive capacities, to nascent systems of indexes. At first these were modest; the systems reflected in advanced lithic technologies need to be thought of in this way. But as the interrelations of the indexes in these systems grew more intricate and complicated, the stage I have termed *hyperindexicality* was attained—and never completely superseded, for it remains a fundamental, varied dimension of human culture today. Only once hyperindexical systems were within reach could fully developed symbolism emerge, but beyond this point symbolic systems must have come frequently to be attendant on hyperindexical sapient cultures. With the coalescing of *indexical* systems, in other words, the slow, sporadic increase in hominin cultural complexity gave rise to a new, emergent force, and *symbolic* systems came to be an attractor in the search space of human cultural possibilities.

Nonsymbolic Neandertals and sapients

While it seems evident that sapient and Neandertal cultures 200,000 years ago were hyperindexical in some degree, I have already suggested that neither was yet symbolic. This is, of course, a matter of interpretation, and it is important to be clear about the criteria for making the call. In itself, ocher body painting, for example—even a transmitted tradition of it—does not much aid my interpretation, and neither does it cinch any argument for symbolism, for, depending on the cultural system of which it formed a part, it could represent a nonsystematic index repeatedly deployed or a systematic one or a full-fledged symbol. The same is true for the other early artifacts that archaeologists habitually label "symbolic," such as beads and incised ocher or other materials. These could all be the products of an advanced indexical culture, and we need to contextualize them by invoking other kinds of evidence.

Carefully studied lithic technologies are central here, and the state of these in the Levallois industry (well established 200,000–300,000 years ago) reveals that the humans practicing them were the possessors and makers of cultural systems of unprecedented sophistication. It is especially the systematization of these practices that convinces me of the incipient hyperindexicality of the cultures in which they were embedded. Even so, other lines of evidence militate against the idea that these were fully symbolic cultures. The somewhat static, unchanging nature of the Neandertal behavioral repertory across the history of the species has often been cited as evidence against their possession of modern symbolic and linguistic capacities, which (the reasoning goes) would have led to more dynamic cultural development. While this position has been nuanced in recent views of Neandertals (Langley, Clarkson, and Ulm 2008; Nowell 2010), there is no reason to discard it completely. We saw in chapter 6 that extraordinary behaviors occasionally appear in the Neandertal record: burials of the dead, hash marks scratched on a cave wall, and an instance of a deep-cave, structured arrangement of broken stalagmites. But these could tally well with an episodic hypertrophy of indexical culture—indeed better, in their sporadic rarity, than with fully symbolic culture.

The behavioral panoply of the first *Homo sapiens* was not much different from Neandertals', and the African evidence discussed in chapter 6 does not suggest any explosive cultural developments from 200,000 to 100,000 years ago. Here too the picture is one of slow, episodic change, though the picture in this case would eventually be transformed. And another question arises in both the sapient and Neandertal cases. If these were fully symbolic species 200,000 years ago, endowed with modern language and the presumed advantages it confers in making culture, shaping taskscapes, and constructing niches, why were they confined to their respective African and Eurasian habitats? Appreciating the immense skill and canny versatility of these early humans does not countervail the evidence of their limited movements and restricted habitats, which were not decisively diversified by either species even in the relatively warm Pleistocene climates from 125,000 to 75,000 years ago—generally *less* challenging than later ones. There is only one "explosion" in all of Paleolithic history that suggests the full powers of linguistic and symbolic culture at work: the permanent expansion of our founding lineage across the world after 70,000 years ago. To understand Neandertals or the first *sapiens*, we should replace the supposition of symbolism with a clearer understanding of the dimensions of hyperindexical culture. One way to do this is to consider the origins of ritual.

: 3 :

The emergence of ritual

What would ritual be like in a human society before language? The question seems counterintuitive, even far-fetched, but it is pressed on us by the picture I have drawn of Middle Paleolithic and Middle Stone Age societies with highly developed indexicality, even hyperindexicality, but without symbolism. Or maybe that picture suggests something more, that the question is another artifact of linguocentrism, for what has led us to think that the rise of human rituals depended on modern language at all? We should ask instead: How could a hyperindexical society *not* perform rituals?

In chapter 5 I proposed that ritual emerged as a concomitant of systems of indexes. When these reached a certain degree of complexity, they took on a holistic identity that made possible their metafunctional operation on the taskscape—their interventions, in their cultural settings, as complete systems. With this holism came also a repeatability of the systems as such, formalized to a novel degree, and this tended to separate off the congeries of activities that composed these systems from other activities. A division arose between the everyday making of the taskscape and another, distinct sphere of its operation. All these features—holism, formal definition, repeatability, and distinctness from surrounding behaviors—are ones anthropologists have found to define the arena of ritual (Tambiah 1979; Bell 2009).

Michael Silverstein (2003, 203), who pointed us toward the indexical nature of ritual, named *indexical iconicity* as its characteristic means of defining and authorizing itself. This concept concerns, for him, the repeatability that enables the organized indexes to ground and guide social discourse and so to form the framework for the dimension of discourse he calls metapragmatics. For us the indexical iconicity of ritual points to the broader semiotic condition that I discussed in chapter 4: the dependency of indexes on iconicity by virtue of their correspondence to similar situations learned earlier. The indexical iconicity of ritual is the systematizing of an aspect basic to indexicality itself, such that the whole organization of activities in the ritual (the system) assumes through iconic repetition the identity of an index (the metafunction). Such metafunction must have begun to take hold as soon as indexical systems arose, before modern language existed; and it carried within it the seeds of ritual, which grew as the systems gained in formal definition and distinctiveness.

From the systematized indexes, also, emerged the original meanings of ritual. For anthropologist Stanley Tambiah (1979), offering a general "performative" theory of human ritual today, indexicality forms one side of the bilateral meaning that all rituals generate. This meaning is divided between content and form. On the one side is a symbolic or iconic content, a representation of a cosmos or worldview that ritual presents in its enactment, and on the other is an indexical form shaped in the enactment itself, which indicates the social positioning and powers of its actors. In modern-day ritual Tambiah gives priority to the symbolic and iconic representation, but from a Middle Paleolithic, forward-looking perspective we can see something different. The earliest moves toward ritual must have been indexical systems that affirmed orders of status and power in hominin groups; these systems intervened in social hierarchies that (to judge from today's nonhuman primates) were in place long before the systems took clear form. In imagining this nascent ritual, we do not need to choose between form and content, for the indexical structure itself *was* the content, expressed as the social orders affirmed through repeated performance: a true aboriginal ritual for an Indexical Age.

Ritual on the taskscape The separateness and holism of nascent ritual are related to the other effects of independence that emerged from complex cultural systems: their autonomy, their attendancy on certain conditions, and their epicyclic impact on the feedback of biocultural history; I will return to these. What set ritual somewhat apart, however, was a double character that it manifested in its repeatable performances, which combined a distinctness from and intimacy with the materials and rhythms of the taskscape. Ritual analyst Catherine Bell describes the process of ritualization as the invoking of "a series of privileged oppositions that, when acted in space and time through a series of movements, gestures, and sounds, effectively structure and nuance an environment" (2009, 140). In nascent ritual, the iteration of activities opposed and yet akin to the daily making of the Paleolithic taskscape introduced to it something new, an additional fold in its formation that arose from the complexities of emerging cultural systems.

Such iteration was also the source of the historicity that ritual created, the sense of *re*enactment in the here and now that yielded a parallel sense of *pre*enactment of, and hence some control over, the future. Tim Ingold has emphasized the temporal aspects of the taskscape, in which past and future are gathered together in a present node of activities, and for him ritual is not radically distinct from these activities. Through networked, embodied social actions the ceremonies of ritual represent the construction of experienced time that is integral to the taskscape; here Ingold takes issue with Durkheim's view of ceremony as a chronology imposed from without in the manner of "guidelines" segmenting societal time through their "periodical recurrence" (Ingold 2000, 196–97). But while it seems right to think with Ingold that the construction of temporality saturates, defines, and even arises from taskscapes—and right to think, additionally, that it has done so since far back in the Middle Pleistocene (in the taskscape at Boxgrove, the temporality was not yet ritualized but was created instead as collaborative urgency or loosely repeated uses of the landscape)—the historicity I propose as a trait emergent in Paleolithic ritual is distinguishable from this socially constructed time. It is a special case of it, which arose in a distinct sphere from a systematization of gestures much exceeding the quotidian. Ritual depended on systems of indexes immanent in the taskscape, and the effects it imparted came from the space they carved out.

By the same token, because ritual historicity is a special kind of the temporal fashioning that pervades the taskscape, we cannot draw a sharp borderline between ritual acts and nonritual (but habitual) activities. The difference will always depend on degrees of systematization. An example will illustrate this. We can picture the pedagogy needed to convey a complex technological sequence from one Neandertal generation to the next as a one-on-one sharing of attention of the sort I described in chapter 4, involving a skilled toolmaker and observant novice. If we extend this only a little, we can picture a group of novices watching and learning from the toolmaker; there is nothing we know about Neandertal society that would render implausible such a scene. Already in the group scene, however, we have shifted to a higher degree of social organization, with perhaps a clearer distinction of it from other activities around it and a clearer marking of the special status of the teacher. A further shift, slight but significant, might see a glimmering awareness of the special import of the group pedagogy, marked by distinct social practices—the decorating of the novices with body paint, for example, or the burning of meat or leaves. Now a metafunction of the cultural system is forming, and with it the special arena of ritual comes into view. But there is no dramatic disjunction to be discovered between ritual and non- or preritualized systems of activities.

This example is meant to suggest also that ritual required no modern language for its coalescence and impact, but only the attainment of a certain degree of complexity in the systematization of indexical culture and protodiscourse. We must work to imagine hyperindexical human societies with nascent ritual but without words and syntax—or, to turn around the proposition and once again evade linguocentrism, we must realize that the coalescing of indexes into systems during the Indexical Age would have pointed culture toward the autonomy and iteration of some of those systems, and thus to their ritual demarcation and historicity. This is another instance of systems forming powerful attractors on the morphospace of Paleolithic cultural possibilities. Again: How could a hyperindexical society *not* perform rituals?

Once the indexical origin of ritual is appreciated, some of the perplexing wonders of ancient human behaviors begin to fade into ancient common practice. Few archaeologists propose that Neandertals possessed a fully modern language, and this negative consensus seems a corroboration to some who doubt the evidence we have that they buried their dead, painted their bodies, etched walls, and found their way deep into caves to make stalagmite arrangements. Without language, these minimalists ask, how could Neandertals attain such cultural sophistication? But if we again reverse the frame and look at the matter prospectively, we can see that Neandertal technologies reveal the advanced cultural systems they mustered, and we can speculate as to the Neandertal modes of communication, sprung from a long selective history enhancing the use of indexes, that ushered such systems also into protodiscourse. Without modern language, Neandertals were nonetheless at the cusp of hyperindexical society and the ritual behaviors it fostered, and there is no reason to suppose that arranging stalagmites and burying their dead—or indeed other, more elaborate rituals—were beyond them

Homo sapiens, emerging from their predecessors by incremental changes, created cultures that were generally similar to these Neandertal ones. This proposes a starting point for the sapient career in somewhat more specific terms than I offered in chapter 6. Early sapient cultures were nonlinguistic and nonsymbolic, but they had systematized the deployment of indexes to a considerable degree, and so were beginning to consolidate their sociality in system and historicity. The resulting iteration of systems of indexical gestures on the taskscape was the minimal condition for their emergent ritual, and the distinctive space of ritual must have formed a strong attractor in these circumstances.

Music, dance, and ritual

Such gestures, Ingold and Bell remind us, cannot be detached from their environmental circumstances. Nascent ritual organized the taskscape in novel ways through repeated, formal movements of bodies and voices interacting with each other and their material surroundings. But while these joint movements and vocalizations were never disembodied or dematerialized, they grew more distinct and autonomous from everyday activities along with the systems of which they were parts. They came to be distinct *choreographies* of bodies, to use a word archaeologist Gamble (2012) has applied, as well as *choruses* of raised voices—useful terms, as long as we do not limit them to contemporary usage but extend them to embrace all manner of more-than-ordinary gesture: rhythmically regular vocal articulations, intonation, chant, song, patterned movements of all sorts, and emergent synchronies among these.

In such movements of body and voice we discover the source of the ubiquitous connection of music and dance to ritual (Tambiah 1979; for more recent examples, see Nettl 2000; Morley 2009; and Brown and Jordania 2011). In *A Million Years of Music* I argued that this connection was aboriginal, an emergence of music from certain kinds of making of the taskscape. The motions of music and dance were at first nothing more than the systematized indexes gathered and made distinct in early ritual, channeled as bodily movement, vocalization, and instrumental prosthesis. This is what encouraged me in that book to define all music today as "the extended, spectacularly formalized, and complexly perceived systematization of ancient, indexical gesture-calls" (2015, 205). The transcendentalism of music, in its earliest form, was never the otherworldly flight that modern ideologies portray but was instead the displacement enacted in the ritual sphere through sounding, patterned, and repeated movement, a special aspect of Maurice Bloch's (2013) transcendental sociality. It was in this way, like ritual itself, an *immanent abstraction* of cultural systems.

This does not imply that Neandertals or early *sapiens* made music in its modern forms, or that their nascent rituals were as fully elaborated as modern rituals typically are. The roots of the abiding coalition of ritual, music, and dance reach down to a period when their modern forms had not taken shape. The deep-seated connection of music and ritual should also not be mistaken for the idea that music came first, language following from it—a model of language origins, repeated and elaborated from Darwin's time to our own, that I opposed in *A Million Years of Music*. Music and language emerged, along their own paths, as parallel outgrowths of the hyperindexicality that culminated the Indexical Age.

This account of nascent ritual suggests that its indexical systems were bifurcated in their impact. On the one hand, they organized movement and vocalization in ways that would drift, finally, toward the formations we call dance and music. On the other, they formed a nodal point of pasts and futures, closing temporal distances and yielding an effect of the presence of absent things, events, or people. In this bifurcation ritual systems were both proximate and abstract. A move beyond the physical and the sensible toward a *metaphysics* was adumbrated here—though we must carefully qualify that word. On the late Paleolithic taskscape, the transcending of physical presence arose as another immanent abstraction, and it can be usefully imagined as the final extension of the "release from proximity" in hominin perceptions and behaviors that has a history much deeper than *Homo sapiens* alone. I will come back to this materialist metaphysics below.

: 4 :

Attendancy, autonomy, epicycle From chapter 5 on, I have used several terms to name the ways in which the systems of late Paleolithic culture grew strong in their social contexts. Their *holism* connotes their tendency, once they had coalesced, to move forward in cultural transmission *as* systems; as they did so, they came naturally to be marked as separate and distinct, *abstracted* from their surroundings. Nascent ritual and its concomitants are special instances of this tendency, creating their performative arena through their organized, repeated gestures on the taskscape. The relations of the systems to their component elements varied according to the natures of these elements—material, behavioral, and ideational—which were all brought into the workings of culture by semiotic connections. These relations could be close and causal, such that, given a certain set of components, a certain system was likely to arise. This is the relation of *attendancy* that I am arguing was responsible for many convergent developments in the biocultural evolution of independent human populations. However, the causality of attendant systems was not a one-way street, and once the congeries of elements caused the attendant system to appear, the system exerted its *autonomous* impact back on the culture of which it was a part. If the bead-making system found in many early human cultures was attendant on its constituent parts (shells, teeth, etc., and ideas and practices of social difference), it also enforced a new way of experiencing those elements after it had formed. Shells were no longer merely shells but potential markers, and social difference was now linked to a particular regimenting of material substance.

Epicycles and prior constraints The hypertrophy of this kind of autonomous operation led to cultural *epicycles*. These are the systems that broke loose from their cultures, so to speak, to achieve an operation so independent that they could turn back and guide culture in the manner of a feed-forward control mechanism. Yet another graded spectrum of possibilities is revealed here, with autonomous feedback shading into epicyclic control, and we might think of epicycles simply as systems that came to form a stronger attractor than others in the cultural search space. But with this special term I want to mark something more than this, a feature not present in more modestly autonomous systems: at least one defining component of an epicycle operates according to a dynamic entirely beyond or prior to human command or alteration, and this forms a driving force of its extended autonomy.

An example concerns the parallel pathways from protodiscourse toward modern language and music, and in particular the shifts in each from graded, analog communicative spectra to discrete, digital ones. These result from two distinct but interacting epicycles. With increasing social and communicative complexity came the proliferation of diverse vocalizations, and as the advantage of these became prominent on the taskscape, selective pressure arose for sharper distinctions in the intonations and timbres the hominin brain could perceive and the hominin vocal tract could produce (Morley 2013). In innate calls and in early protodiscourse, intonational shapes and different timbres had occupied positions along an unbroken continuum of vocal production. These are the communicative gestures anthropologist Robbins Burling (1993, 2007) has termed "gesture-calls," and their analog nature persists today in our innate vocal and bodily gestures: in the continuous gradations of emotional expressions either vocalized (sighs, sobs, laughter, and more) or embodied (smiles, frowns, shrugs, and the like). The offloading of the calls to culture and the resulting proliferation of vocalizations, however, crowded these analog signals together on their continuous spectra, and this created pressure for modes of discreteness that could maintain each signal's distinctness. A selective bias arose that was accommodated through the same new cognitive capacities that enabled the multiplication of vocal signs and their interpretation, forming a feedback between signal demands and signal production and perception.

The resulting shift toward discreteness itself altered the selective environment, as the advantage of communicative precision and variety created its own niche-constructive loops. Under the influence of the different social uses to which vocalization was put in hyperindexical societies, discreteness took hold in different ways. On the one hand was the discretization of timbre or tone color, retained today in the small arrays of distinct vowel sonorities that characterize the phonemic repertories of all languages; on the other was the focusing of frequency processing that created the perception of discrete pitch levels, carved out of the smooth intonational contours of the earlier calls. These two modes of discreteness were overlaid on top of the continuous spectra of protodiscourse, redoubling communicative complexities and adding new kinds of elements (and percepts) to them. They led to combinatorial systems: the phonologies of modern language—discretized phonemes from which morphemes and words are built—and the arrays of discretized pitches that serve as the gravitational centers of melodies in music (Tomlinson 2015).

Of the elements composing these two epicycles, some were immanent in the cultures in which they arose, some were determined by the actors in those cultures, and most were shaped and changed by niche-constructive feedback across biocultural history. But what determined the powerful, guiding control of these particular histories—what made them epicycles—

was a dynamic that was ontologically prior to all this: the fact that multiplying analog signals and crowding them together on a continuous spectrum will ultimately undermine the distinctness of similar signals. This is more basic even than semiosis itself, a constraint of the informational realm: wherever discreteness of signal is lacking, similarity of signals courts ambiguity. In late indexical cultures, this constraint moved to the heart of cultural systems of communication, and when this occurred, epicycles driving toward discreteness took shape.

It is important to be clear about the largest loops of feedback and feedforward involved here. The state of indexical use in these cultures brought a prior, ontological constraint into play in the feedback relations of humans and their niches, with the result that the systems were pulled into a feedforward, controlling alignment with respect to the biocultural history of which they were a part. In defining epicycles at the end of chapter 5, I maintained that the involvement of such prior conditions is a basic aspect of their formation, cultural or otherwise. The move toward discrete combinatoriality in human communication systems offers an example from the cultural realm.

An epicycle of linguistic syntax Another epicycle determined the coalescence of the syntax and grammar of modern language and has been described by Terrence Deacon. Deacon's model of language origins is the most compelling of the many that have been offered in recent years, largely because of the two strains of thought he joins in it. The first is semiotics (he has led the way in reinstating Peirce's views in the field of language cognition); the second is niche construction (Deacon 1997, 2003b, 2010, 2012a). For all its strengths, however, Deacon's model is an avowedly symbolocentric one, and some of its truest advantages are obscured by the preeminence he accords symbolic cognition in human evolution. To see clearly these advantages requires revision of the model and some weaning of it from its symbolic emphases.

As I described in chapter 4, Deacon has elaborated Peirce's discovery of the hierarchic relations among sign types, in which indexes are dependent on iconism for their meanings, and symbols in turn are dependent on indexicality. These relations must be understood in the light of an interpretive process of semiosis rather than a typology of signs or a static relation between sign and object, and this process is the operation of Peirce's interpretant. For symbols, the nature of the relation is particularly complex (Deacon 2012a). A symbol such as a word in a language derives its referential power not from an arbitrary conveyance of meaning in the manner of a code or token but from its functioning in a system. The nature of this referential system is defined by three general semiotic aspects or constraints:

an array of signs internally differentiated one from another; a functioning of the system as a whole as an index pointing to another array of objects (or actions, ideas, etc.); and a fixing of the relations among the signs within the system such that one can point indexically to another.

The third aspect, Deacon argues (2012a), not only creates symbols but is the source of syntax and grammar. The structure of language propositions, in which the meaning of a predicate (typically a verb or verb phrase) is filled out by one or more arguments (usually a noun, noun phrase, or pronoun), corresponds to the semiotic process by which the reference of one sign (it will become the symbol) is specified and completed by the pointing indication of another (hence an index). This connection of linguistic syntax to the semiotic condition of symbolism is revealed in the many deictic functions of language that attach arguments, and thereby their predicates, to the worldly contexts of language users: pronouns pointing to nouns ("he," "you," "which"), indicators ("this," "that," "here," "there," "next," "last"), quantifiers ("some," "many," "all"), possessors ("hers," "yours"), and more. These linguistic functions are indexes, pointing the predication (verb and nouns together) of a proposition to things and situations in the world. Without them words have sense—in the manner of an arbitrary code, again—but they do not refer to specific, real-world contexts. Words depend on indexical cues to reach out beyond their system to the world in the moment-to-moment flux of communication. (I consider here only the referentiality brought about within the syntactic system itself; there are also pragmatic and metapragmatic indexes, pointers generated in the contexts in which language is deployed that pin speech to real-world situations. The roots of these reach deep into the presymbolic ordering of gesture-calls and Bowie's protodiscourse; see Tomlinson 2015.) Grammar and syntax, Deacon argues, are not externally given laws but "intrinsic symbolic attributes that emerge into relevance as symbols are brought into various semiotic relationships with one another"—relationships that involve the indexicality on which symbolism is founded (2012a, 17). In other words, grammar and syntax emerge from the complex hierarchization that underlies semiosis of the symbolic kind, which always involves indexes as well as symbols.

From hyperindexicality to symbolism

This semiotic constraint, like the informational one I described above in the phonological and discrete-pitch epicycles, is a prior condition arising from the hierarchic relation of sign types. It lies beyond any reshaping or contravening in cultural systems themselves, and it suggests that Deacon has here modeled a consequential epicycle in the formation of modern language. Extending his argument, we can speculate how this epicycle took off from the hyperindexical cultures of early *Homo sapiens*. As these cultures generated indexes that were more and more distinct, arrays of indexes formed, and in the presence of hierarchizing cognition their internal distinctions fell into incipient ordering. But the systematized indexes could not become symbols without the formation of a further hierarchy, between the nascent systems and arrays of further signs indexically specifying their reference. Indexes were ubiquitous, however, the fundamental semiotic units of hyperindexical culture and advanced protodiscourse, and in this situation an attendancy arose linking the indexes systematized in hierarchical orders to other indexes outside these orders. These links, then, created new, hierarchized ways of extending referential precision—new ways of enhancing the powerful adaptive advantage of all the semiotic richness at hand—and this pushed the systems toward full symbolism. As this occurred, however, the new dimensions of communicative reference had to conform to fundamental semiotic constraints and hierarchies, and so an epicycle molding linguistic predication in the guise of the symbol-index relation took shape.

The nature of the cognition at stake in symbolism and syntax is a major concern of Deacon's (2012a), and he divides up linguistic functions between different brain systems. He aligns the indexical aspects of syntax with procedural memory, which allows for deep automatization and rapid-fire processing, and the symbolic aspects with the slower, more cumbersome, but expandable episodic or associative memory. From a cognitive vantage this seems a convincing proposal, and it answers to the fact that our modern-day processing facility is quicker and more automatic for deictic functions of language than for the complexities of predicates and arguments.

From a deep-historical vantage, we can discern something else in Deacon's proposal: a broad-stroke retracing of the communicative evolution of the Indexical Age. The hominins of the Boxgrove period, I have argued, deployed indexes in complex fashion, and by the time of Neandertals and early sapients this complexity was redoubled. Alongside these semiotic changes must have come changes in hominin brains, as computational modes incapable of much systematization of memory gave way to richer, more hierarchized possibilities (see chapter 5). The correlation of these two histories seems inescapable, and indeed, the enhanced hierarchization required before the emergence of symbolic systems from hyperindexical ones could occur seems to match well the upsurge in hierarchic complexity evident in technological systems among Neandertals and *sapiens*. Note that the archaeological evidence here takes the form, not of single artifacts that we pronounce to be "symbolic," but rather of cognitive patterns that we infer from technological operational sequences and that correspond to the patterns we discern in symbolism. We can in this view, then, specify not only the epicyclic drivers of the emergence of symbolism but also some archaeological proxies that suggest when the cognitive requisites for it were in place. These all point to a period after about 200,000 years ago as the era on the verge of transforming some aspects of hyperindexicality into symbolism.

Deacon would disagree, arguing that symbolism reaches back as far as a million years, deep into the era of *Homo ergaster* and even before *Homo heidelbergensis*, the presumed ancestor of Neandertals and *sapiens*, appeared. But here his symbolocentrism has misled him, for the implausibility of his timeline is clear both on archaeological and on evolutionary grounds. From the archaeological side, we have no evidence of complexly hierarchized technological operations for the first half of this period: it was still the era of Acheulean hand axes, modestly hierarchized at most, and it is wishful to perceive in their operational sequences traces of the complex cognition needed for symbolism. Indeed, the kinds of artifacts that might persuade us we should be looking for such cognitive complexity are lacking until long after the 500,000-year mark. Moreover, the attainment of symbolism and modern language, if it is as powerful a force as Deacon believes (and as most, myself included, agree), should have transformed culture and society, but such transformation is nowhere to be found in such behavioral patterns of Acheulean hominins as we can reconstruct. Its absence, I have noted, continues as late as the Neandertals to pose a stumbling block for proposals of modern linguistic and symbolic capacities.

On the evolutionary side, Deacon's argument is that the brain systems required for symbolic cognition are too many and too complexly aligned for evolution to have built them over the 200,000-year span of *Homo sapiens*; something like a million years is needed. But his own evolutionary evidence does not support so lengthy a chronology. First, Deacon argues that symbolism and language were emergent phenomena, arising in "neural structures and circuits . . . found ubiquitously in most monkey and ape brains"—"old structures performing new tricks," as he puts it (2012a, 34). If these circuits were widely dispersed and in place long before symbolism or *sapiens* arose, however, this suggests that they did not require wholesale overhaul in order for the symbolic emergence to happen, but only the kind of incremental changes in cognitive patterns ("tweaks") that I described in chapter 6. For symbolism this no doubt involved, as Deacon correctly maintains, plastic rerouting and new recruitments of brain systems—complex shifts in ontogenetic development as well as phylogeny. But he marshals no clear evidence that it required major rebuilding of wetware.

The impact of niche construction

The emergence of symbolism required also the dynamics of cultural niche construction, and it is one of the chief advantages of Deacon's account that he recognizes this force in human evolution. Here too, however, his own evidence weighs in against his long timeline. His chief example of the offloading of innate behaviors to social transmission—of niche-constructive buffering, in other words—is the domesticated Bengalese finch, which, in comparison to its wild variety, shows a much broadened range of song types. This diversification, Deacon and others have argued, resulted from domesticated breeding of the birds, which in effect erased song as a factor in their reproductive success. In this artificial selective environment, the innate constraints on song structures were weakened through mutations not selectively culled; then existent brain circuits were recruited for new functions of song transmission and learning—cultural functions, in the broad sense I have advocated—and the variety of song blossomed (Deacon 2010, 2012a). The hypothesis seems convincing, and its mechanism is certainly so. But if all this cognitive and behavioral change could come about in finches across a mere 300 years of domestication, how much alteration could cultural niche construction, with its buffering, accelerations, and the rest, bring about in *Homo sapiens* across 150,000? And how much more when we factor in the effects of cultural systems and epicycles?

The problem with symbolo- and linguocentrism is that they lead even the best accounts toward simplification and reduction of the impact of cultural niche construction, when a full account of our modernity requires instead an understanding of niche construction more sophisticated and elaborated than has hitherto been realized.

: 5 :

Techne At base, the dynamic of niche construction is the story of humans scratching out their lifeways on the material world, malleable or intractable, stable or changing, obliging or meager. Matter and the energy to manipulate it formed the most basic constraints and opportunities for Paleolithic humans, and their exploitation of worldly affordances formed their technologies and shaped their taskscapes. The traces we have of this work are material remainders of the mediation between the semiotic components of human cultures and the behaviors that expressed them.

This mediation was not stable across the several million years of hominin evolution, of course, but was marked by small leaps, often repeated, alongside periods of repetition and stasis: the discontinuous gradualism I described in chapter 6. The unevenness is not merely an artifact of the imperfect record we have and not only a reflection of species-wide capacities, but reveals the varying circumstances of local populations as well; it continued on down to Neandertals and *Homo sapiens*. Arching across all this uneven development, however, there is a large, general trajectory: a widening and complicating of the mediated space between raw materials and their manufactured uses. This trajectory is revealed in several kinds of evidence archaeologists have tracked in impressive detail: in the gradual expansion of the series of steps employed in the operational sequences of lithic toolmaking; in the lengthened curation of tools, which came more and more often to be retained and reused, often with retouching and overhaul; and in the longer distances across which raw materials were transported from their harvesting to their use. These signs of widening mediation appear through much of hominin history, but their occurrence accelerated and grew regular among both Neandertals and *sapiens* in the tens of millennia around the Middle/Upper Pleistocene boundary, some 127,000 years ago (Gamble 1999, 2007; White and Ashton 2003; Langley, Clarkson, and Ulm 2008).

Technical abstraction

The distance between matter and its use is a form of abstraction, and each of these three indicators of it shows an enlargement of the off-line thinking that defines Bloch's transcendental sociality discussed in chapter 5—of thought, that is, ranging away from immediacy, social copresence, and things apparent to the senses. Longer operational sequences reflect increasing conceptual distances between raw materials and finished tools, lengthened curation manifests an expanded temporal conception of a tool's use-life, and long-distance transport is the enactment of a broadened imaginary of the taskscape itself. In this kind of thinking there is a foresight and planning that seems to have been novel in degree and consistency in the period of late Neandertals and the species-span of *Homo sapiens*. It is a thinking, moreover, that shows a binding of past and future pointing in the direction of the formalized historicity of ritual.

But what is the force that abstracts thought from presence in these technologies? What is it that mediates? It is the systematizing of culture itself; the widened distances were opened by the swelling complexity built into these systems. The technological manipulation of matter moved along a track parallel to the coalescence of ritual forms and the semiotic shift from indexicality to hyperindexicality and symbolism.

This is most evident in the operational sequences, which achieved new length and intricacy through ramifying systematization, especially the hierarchization of procedure and gesture. This led to recursive technologies of several sorts (d'Errico et al. 2003): techniques that prepared for other techniques (Levallois methods already exploit these), tools made to function as parts of composite tools (the point, shaft, straps, and glue of a spear), and tools made in order to fashion other tools (antler and bone used for fine-grade knapping, stone awls for bead making, and bone needles for sewing clothing). Instances of such recursion no doubt reach far back, but it attained a new prominence, new complication, and new versatility on the technological horizon in the period from 200,000 to 50,000 years ago.

The abstraction reflected in long-distance transport of materials and lengthened curation of tools shows a related extension beyond the present moment of sensed experience, one that relies also on the systematization of thought and behavior. In order to become a regular practice, transport of useful materials across dozens or hundreds of kilometers required either the movement of groups across a hugely expanded taskscape or the equally suggestive intercourse and exchange between groups. In both cases the structuring of societies bore new relations to material affordances, dependent on long-term organization of movements, prior planning for material engagements and harvesting, and, in the case of exchange, novel conceptions of group identity. The result was a shift toward new economies of materials and also, more deeply, toward the "solutions of the problems of absence" that Clive Gamble (2007, 211) has seen as constitutive of modern societies. Extended curation of tools, meanwhile, represented the ushering of factured materials into the *ongoing* fashioning of the taskscape, in contrast to the opportunistic exploitation of materials in most earlier hominin toolmaking and in nonhuman toolmaking today. This was linked to the deepened accumulation of cultural archives that distinguishes late hominin species from other animals and, again, to the abstracted temporality that ultimately emerged from the systems spawned by these archives.

A new conceptual relation to nonpresence thus arose from the abstracting tendencies of cultural systems. Gamble describes this relation, with a phrase borrowed from earlier writers, as a "release from proximity" (Gamble 1999, 2007; see also Rodseth et al. 1991). I elaborated this notion in chapters 4 and 5, where I suggested that there are two distinct modes or phases of the release, one general to all complex animal cultures and the other specific to late hominins. The first is the extension of semiosis in culture, by which signs are displaced from their situations of use and transmitted for use in other contexts. In highly developed form this leads to Bloch's transcendental sociality. The second kind of release—it is, more exactly, an extension of the first in degree—carries this movement of culture farther, into the realm of systematics. The abstraction brought about by growing systematization, which blossoms into all the features I have described, including holism, autonomy, and epicyclic action, is the unique and distinctive feature of late hominin cultures, and its arena reaches beyond ritual, music, and language into the very heart of the technological mediation of materials. In saying before that the systems of culture mediate the space between raw materials and their technical exploitation, I wanted to affirm this general truth; but we can see that the assertion does something more: it links advancing technology to all the modes of abstraction, displacement, and distancing of thought that set late hominin biocultural evolution apart from every other evolutionary development that the earthly biosphere has seen.

Immanent metaphysics

At this material heart of techne there appears, then, another possibility that I signaled above: the seamless extension of the release from proximity toward realms farther afield than resources beyond the horizon or the future use of a tool or an elaborated plan for making it. The possibility arises not merely of the surmounting of copresence in human experience but of a more radical surmounting of sensible experience altogether to broach supersensible worlds. These were never worlds utterly out of touch with the conditions and experience that shaped the human imaginary, for, then as now, we remain, even in our highest imaginative flights, restricted to realms that are not fully but only partially counterintuitive (Tremlin 2006). Nevertheless, the interposing of systems between matter and its uses, the marking off of ritual space, performance, and historicity, the hyperindexicality of music, and the culminating elaborations of semiosis involved in symbolism and language all transcended in their own ways the boundaries that had always before delimited animal experience—including *cultural* animal experience. They uncovered, immanent in systematization itself, not only a movement beyond the physical but also a new force for shaping that beyond in ways that have characterized humanity ever since. To these ways we give many names: magic, animism, spiritualism, divinity, gods, religion, enlightenment, and more. In their discovery we see another extension of Bloch's transcendental sociality, one that reveals, in his words, "why religion is nothing special, but is central" (2008).

In pursuing such thoughts, however, it remains important not to lose sight of the ground below. Metaphysics, in the sense of what lies beyond the physical, arose as another immanent abstraction in human culture. Its appearance was not a simple matter of evolved modules for animal behavior inevitably determining the presence of gods, as many accounts have argued (see Boyer 2001; Atran 2002; Tremlin 2006; Barrett 2011), and neither was it an untethered leap into the immaterial beyond. Instead, it was an outgrowth of the experience of matter itself, under the aegis of sapient computational cognition and especially under the impact of cultural systems. The distance between an imaginary that sponsored the carrying of raw materials over the hill to alter and exploit them on the other side and an imaginary that could fashion depictions of figures that never were, like the therianthropic lion-man of Hohlenstein-Stadel (Conard et al. 2015), was a distance filled in, across sapient populations from South Africa to France to Australia, with technologies marking many lengths and degrees of intricacy of system. I have argued that it was a distance filled in also by other fundamental cultural systems, not specifically and narrowly technological but communicative and defining new social dimensions. "Filling in," however, is not quite the right verb: instead, the distance was opened out, articulated, and *constructed* by the systems themselves. It was these that reflected new cognitive possibilities and, in doing so, introduced new ways of niche making, these that brought to light possibilities of transcending materiality through material manipulation itself.

The release from proximity, however far it reached, was only ever a continuous if uneven terrain of cultural production. It emerged from a building of niches in which certain outcomes came to be more likely than others because of the abstract machines at work in them, and it resulted in a search space of biocultural possibilities where metaphysics came to be an inescapable, powerful attractor.

: 6 :

The substance of my argument in this book has been that modernity came late and quickly to *Homo sapiens*, but that this development requires for its explanation only the persistent, potent dynamics of niche construction interacting with cultural systems of increasing elaboration. The extraordinary force of this coalition between a niche making as old as life itself and cultural systems new and unique to hominins (in complex form, to late hominins) reshaped natural selection and made us who we are; the process, however long its anticipations and however many its sporadic head starts, accelerated and culminated in the first 150,000 years or so of sapient evolution. All the models evolutionists offer suggest that such large changes fall within the range of potential effects of natural selection interacting with cultural niche construction; all the evidence archaeologists have amassed points to this period as one of accelerating and finally pivotal change.

Near-modern and modern

To say that human modernity burst upon the scene within the species-span of *Homo sapiens*, however, is to open the door to a reductive summary that will portray the developments I have described as a leap rather than a process and misrepresent the profile I have drawn of the near-modern natures of Neandertals and the earliest sapients. My model does not cast in doubt the behavioral and communicative sophistication of the humans who lived before and even alongside the coalescing of niche construction and advanced cultural systems. They enacted cultures of a depth that had not been seen before in earthly life; they communicated using an indexical versatility also unprecedented; their technological workings with the materials of the world beggared those of earlier hominins in intricacy, systematic planning, and learned and transmitted content. All these things enabled them to carve out new kinds of lifeways to meet the challenges of Middle and Upper Pleistocene existence. The arguments that persist today among archaeologists and others about human modernity too often rely on unnuanced dichotomies, as if we had never left behind the kind of depictions of Neandertals and "cavemen" that were the norm a century ago. Sharp lines are drawn between talking humans and mute brutes, symbolic "artists" and inexpressive flint knappers, canny hunters and builders and scavengers driven by little more than instinct and fear. Such dichotomies ought to be far in our past, since we have long been in a position to appreciate the gradations of complexity that connect their polar options. It is not a belittling of Neandertal or early sapient achievement to recognize a difference in later achievement that transformed the world, and here it is worth remembering my picture of complexly indexical and even hyperindexical, ritualized societies thriving without fully modern language or symbolic cognition. Neither does it denigrate those earlier humans to propose that the main ingredients of the transformation—hypertrophied technology, fully developed language, an expanded release from proximity and its extension into metaphysics, modern rites and their musical and bodily appurtenances—were barely nascent among them. We can affirm this, on the basis of many kinds of evidence, at the same time as we reconstruct, analyze, and indeed are moved by the immense social sophistication and behavioral sharp-wittedness they managed.

System, selection, and feedback

Nevertheless, the impact of niche construction was redoubled and more by its amalgamation with deep, intricate cultural systems. Systematicity of culture was not new to *Homo sapiens*, having roots reaching far back in hominin history, but the systems now came to embody unprecedented degrees of complexity. Involved here is a shift in cognitive capacity that remains opaque to the evidence we have—perhaps, as I and many others have suggested, it was a marker of sapient difference already 200,000 years ago—but it is important to recall that the modeling of computational machines suggests that the shift need not have been large. To deepen the sedimented cultural archives, already deep among Neandertals and near-sapient humans, the cognitive tweak required only slightly enhanced capacities for storing the transmitted bodies of cultural semiosis. This deepening then offered advantages in niche making that could become targets of selection. The increased depth of cultural archives could not be maintained without ordering their materials; and so the hierarchizing of cultural knowledge (also nascent in earlier cultures) was brought, more powerfully than before, into the play of selection. Feedback mechanisms were activated in new directions, and the structured, organizing force of culture came to have an impact on these mechanisms that was previously unknown.

The feedback cycles drove the enhancement of capacities to make cultural systems more and more complex, and these systems led to the many features I have elaborated: attendancy and convergence, holism, autonomy, and epicyclic feedforward control. But this was not a dynamic without end. Natural selection for enhanced cultural capacities could gain traction only through the *insufficiency* of culture—only as long, that is, as culture was an inadequate mechanism in buffering humans from the vicissitudes of their niches. At a certain threshold point, culture, or, more exactly, the capacity to elaborate it, loosed itself from the niche-constructive feedback mechanisms into which it had entered with new potency 150 millennia before. At this moment a modernity was achieved that has, in its broadest, most basic aspects, never been superseded. The moment is revealed to us not in an Upper Paleolithic, cave-painting revolution or in the shifting of taskscapes toward sedentism and agriculture, the markers usually associated with the dawning Neolithic age; and it certainly did not await an attainment within the last several millennia of moral probity or spiritual wisdom, as advocates of an "Axial Age" would have it. Instead, the moment was marked by the ability of a small lineage within *Homo sapiens*, struggling survivors of the most recent climatic downturn, to fan out through Africa and across the world starting more than 60,000 years ago. Now the hominin niche expanded quickly until it commanded every landmass but Antarctica. Now, with consequences whose fatefulness we still do not properly heed, our taskscape became the planet itself.

ACKNOWLEDGMENTS

I am grateful to many scholars with whom I have interacted as this work has taken shape, among them Carolyn Abbate, Dipesh Chakrabarty, Michele Cometa, Nick Conard, Terry Deacon, Paul Kockelman, Sally McBrearty, Mihaela Pavličev, Rick Prum, Dan Smail, Henry Staten, and Kim Sterelny, and also to the readers at the University of Chicago Press, whose responses helped me to strengthen the final draft. Closer to home, friends, colleagues, and students have had a hand in shaping my thought; I single out three whose colloquy with me has been especially frequent and influential: Francesco Casetti, Günter Wagner, and Juliet Fleming. My children were on my mind as I wrote, and their conversation also has been invaluable.

I offer this book in loving dedication to Juliet.

WORKS CITED

Aiello, Leslie C., and R. I. M. Dunbar. 1993. "Neocortex Size, Group Size, and the Evolution of Language." *Current Anthropology* 34:184–93.

Aiello, Leslie C., and Peter Wheeler. 1995. "The Expensive Tissue Hypothesis: The Brain and the Digestive System in Human and Primate Evolution." *Current Anthropology* 36:199–221.

Alcock, John. 2001. *The Triumph of Sociobiology*. Oxford: Oxford University Press.

Ambrose, S. H. 1998. "Late Pleistocene Human Population Bottlenecks, Volcanic Winter, and the Differentiation of Modern Humans." *Journal of Human Evolution* 34:623–51.

Atkin, Albert. 2013. "Peirce's Theory of Signs." In *The Stanford Encyclopedia of Philosophy*, Summer 2013 ed., edited by Edward N. Zalta. http://plato.stanford.edu/archives/sum2013/entries/peirce-semiotics/.

Atkinson, Quentin D., Russell D. Gray, and Alexei J. Drummond. 2008. "mtDNA Variation Predicts Population Size in Humans and Reveals a Major Southern Asian Chapter in Human Prehistory." *Molecular Biology and Evolution* 25: 468–74.

———. 2009. "Bayesian Coalescent Inference of Major Human Mitochondrial DNA Haplogroup Expansions in Africa." *Proceedings of the Royal Society B* 276: 367–73.

Atran, Scott. 2002. *In Gods We Trust: The Evolutionary Landscape of Religion*. Oxford: Oxford University Press.

Balari, Sergio, and Guillermo Lorenzo. 2013. *Computational Phenotypes: Towards an Evolutionary Developmental Biolinguistics*. Oxford: Oxford University Press.

Barham, Lawrence. 2007. "Modern Is as Modern Does? Technological Trends and Thresholds in the South-Central African Record." In Mellars et al. 2007, 165–76.

Barkow, Jerome H., Leda Cosmides, and John Tooby. 1992. *The Adapted Mind: Evolutionary Psychology and the Generation of Culture*. Oxford: Oxford University Press.

Barrett, Justin L. 2004. *Why Would Anyone Believe in God?* Lanham, MD: AltaMira Press.

———. 2011. *Cognitive Science, Religion, and Theology: From Human Minds to Divine Minds*. West Conshohocken, PA: Templeton.

Barton, R. Nick E., Abdeljalil Bouzouggar, Christopher Bronk Ramsey, Simon Collcutt, Thomas F. G. Higham, Louise T. Humphrey, Simon Parfitt, et al. 2007. "Abrupt Climatic Change and Chronology of the Upper Palaeolithic in Northern and Eastern Morocco." In Mellars et al. 2007, 177–86.

Bar-Yosef, Ofer. 2007. "The Dispersal of Modern Humans in Eurasia: A Cultural Interpretation." In Mellars et al. 2007, 207–17.

Bechtel, William, and Robert C. Richardson. 2010. *Discovering Complexity: Decomposition and Localization as Strategies in Scientific Research*. Cambridge, MA: MIT Press.

Bedau, Mark. 2002. "Downward Causation and the Autonomy of Weak Emergence." *Principia* 6:5–50.

Bell, Catherine. 2009. *Ritual Theory, Ritual Practice*. Oxford: Oxford University Press.

Bertalanffy, Ludwig von. 1969. *General System Theory: Foundations, Development, Applications*. New York: Braziller.

Beyin, Ammanuel. 2011. "Upper Pleistocene Human Dispersals out of Africa: A Review of the Current State of the Debate." *International Journal of Evolutionary Biology*, 2011, Article ID 615094.

Bickerton, Derek. 1990. *Language and Species*. Chicago: University of Chicago Press.

———. 2009. *Adam's Tongue: How Humans Made Language, How Language Made Humans*. New York: Hill and Wang.

Bloch, Maurice. 2000. "A Well-Disposed Social Anthropologist's Problem with Memes." In *Darwinizing Culture: The Status of Memetics as a Science*, edited by Robert Aunger, 189–203. Oxford: Oxford University Press.

———. 2008. "Why Religion Is Nothing Special, but Is Central." *Philosophical Transactions of the Royal Society B* 363:2055–61.

———. 2013. *In and out of Each Other's Bodies: Theory of Mind, Evolution, Truth, and the Nature of the Social*. Boulder, CO: Paradigm.

Bluff, Lucas A., Alex A. S. Weir, Christian Rutz, Johanna H. Wimpenny, and Alex Kacelnik. 2007. "Tool Related Cognition in New Caledonian Crows." *Comparative Cognition and Behavior Reviews* 2:1–25.

Boëda, Eric. 1995. "Levallois: A Volumetric Construction, Methods, a Technique." In Dibble and Bar-Josef 1995, 41–68.

Bowie, Jill. 2008. "Proto-discourse and the Emergence of Compositionality." *Interaction Studies* 9:18–33.

Boyd, Robert, and Peter J. Richerson. 1985. *Culture and the Evolutionary Process*. Chicago: University of Chicago Press.

———. 2005. *The Origin and Evolution of Cultures*. New York: Oxford University Press.

Boyer, Pascal. 2001. *Religion Explained: The Evolutionary Origins of Religious Thought*. New York: Basic Books.

Bradtmöller, Marcel, Andreas Pastoors, Bernhard Weninger, and Gerd-Christian Weniger. 2010. "The Repeated Replacement Model—Rapid Climate Change and Population Dynamics in Late Pleistocene Europe." *Quaternary International* 247:38–49.

Brown, Stephen, and Joseph Jordania. 2011. "Universals in the World's Musics." *Psychology of Music* 41:229–48.

Buisson, Dominique. 1990. "Les flûtes paléolithiques d'Isturitz (Pyrénées-Atlantiques)." *Bulletin de la Société préhistorique française* 87:420–33.

Burling, Robbins. 1993. "Primate Calls, Human Language, and Nonverbal Communication." *Current Anthropology* 34:25–53.

———. 2007. *The Talking Ape: How Language Evolved*. Oxford: Oxford University Press.

Campbell, D. T. 1974. "Downward Causation in Hierarchically Organized Biological Systems." In *Studies in the Philosophy of Biology: Reduction and Related Problems*, edited by F. J. Ayala and T. Dobzhansky, 179–86. Berkeley: University of California Press.

Carruthers, Peter, and Peter K. Smith. 1996. *Theories of Theories of Mind*. Cambridge: Cambridge University Press.

Cavalli-Sforza, L. Luca, and Marcus W. Feldman. 1981. *Cultural Transmission and Evolution: A Quantitative Approach*. Princeton, NJ: Princeton University Press.

———. 2003. "The Application of Molecular Genetic Approaches to the Study of Human Evolution." *Nature Genetics Supplement* 33:266–75.

Chapais, Bernard. 2008. *Primeval Kinship: How Pair-Bonding Gave Birth to Human Society*. Cambridge, MA: Harvard University Press.

Chappell, J., and A. Kacelnik. 2002. "Tool Selectivity in a Non-mammal, the New Caledonian Crow (*Corvus moneduloides*)." *Animal Cognition* 5:71–78.

Chazan, Michael. 1997. "Redefining Levallois." *Journal of Human Evolution* 33:719–35.

Cheney, Dorothy L., and Robert M. Seyfarth. 2007. *Baboon Metaphysics: The Evolution of a Social Mind*. Chicago: University of Chicago Press.

Chiaroni, Jacques, Peter A. Underhill, and L. Luca Cavalli-Sforza. 2009. "Y Chromosome Diversity, Human Expansion, Drift, and Cultural Evolution." *Proceedings of the National Academy of Sciences* 106:20174–79.

Clarkson, Chris, Zenobia Jacobs, Ben Marwick, Richard Fullagar, Lynley Wallis, Mike Smith, Richard G. Roberts, et al. 2017. "Human Occupation of Northern Australia by 65,000 Years Ago." *Nature*. doi:10.1028/nature22968.

Cochrane, Gregory, and Henry Harpending. 2009. *The 10,000 Year Explosion: How Civilization Accelerated Human Evolution*. New York: Basic Books.

Conard, Nicholas. 2007. "Cultural Evolution in Africa and Eurasia during the Middle and Late Pleistocene." In *Handbook of Paleoanthropology*, edited by Winfried Henke and Ian Tattersall, 3:2001–37. 3 vols. Berlin: Springer Verlag.

Conard, Nicholas J., Michael Bolus, Ewa Dutkiewicz, and Sibylle Wolf. 2015. *Eiszeitarchäologie auf der Schwäbischen Alb*. Tübingen: Kerns Verlag.

Conard, Nicholas J., Maria Malina, and Susanne C. Münzel. 2009. "New Flutes Document the Earliest Musical Tradition in Southwestern Germany." *Nature* 460:737–40.

Coop, Graham, Joseph K. Pickrell, John Novembre, Sridhar Kudaravalli, Jun Li, Devin Abscher, Richard M. Myers, Luigi Luca Cavalli-Sforza, Marcus W. Feldman, and Jonathan K. Pritchard. 2009. "The Role of Geography in Human Adaptation." *PLoS Genetics* 5:1–16.

Cooper, A. C., R. Dean, and Cyril Hinshelwood. 1968. "Factors Affecting the Growth of Bacterial Colonies on Agar Plates." *Proceedings of the Royal Society B*. doi:10.1098/rspb.1968.0063.

Darwin, Charles. (1859) 2003. *The Origin of Species and The Voyage of the* Beagle. New York: Knopf.

———. (1871) 1981. *The Descent of Man, and Selection in Relation to Sex*. Princeton, NJ: Princeton University Press.

Davidson, Ian, and William Noble. 1993. "Tools and Language in Human Evolution." In *Tools, Language and Cognition in Human Evolution*, edited by K. R. Gibson and T. Ingold, 363–88. Cambridge: Cambridge University Press.

Dawkins, Richard. 1976. *The Selfish Gene*. Oxford: Oxford University Press.

———. 1982. *The Extended Phenotype*. Oxford: Oxford University Press.

Deacon, Terrence. 1997. *The Symbolic Species: The Co-evolution of Language and the Brain*. New York: Norton.

———. 1999. "Memes as Signs: The Trouble with Memes (and What to Do about It)." *Semiotic Review of Books* 10:1–3.

———. 2003a. "The Hierarchic Logic of Emergence: Untangling the Interdependence of Evolution and Self-Organization." In *Evolution and Learning: The Baldwin Effect Reconsidered*, edited by Bruce H. Weber and David J. Depew, 273–308. Cambridge, MA: MIT Press.

———. 2003b. "Multilevel Selection in a Complex Adaptive System: The Problem of Language Origins." In *Evolution and Learning: The Baldwin Effect Reconsidered*, edited by Bruce H. Weber and David J. Depew, 81–106. Cambridge, MA: MIT Press.

———. 2007–8. "Shannon-Boltzmann-Darwin: Redefining Information." *Cognitive Semiotics* 1:123–48, 2:169–96.

———. 2010. "A Role for Relaxed Selection in the Evolution of the Language Capacity." *Proceedings of the National Academy of Sciences* 107, supplement 2: 9000–9006.

———. 2012a. "Beyond the Symbolic Species." In *The Symbolic Species Evolved*, edited by Theresa Schilhab, Frederik Stjernfelt, and Terrence Deacon, 9–38. Berlin: Springer Verlag.

———. 2012b. *Incomplete Nature: How Mind Emerged from Matter*. New York: Norton.

———. 2012c. "The Symbol Concept." In *The Oxford Handbook of Language Evolution*, edited by Maggie Tallerman and Kathleen R. Gibson, 393–405. Oxford: Oxford University Press.

De Landa, Manuel. 1997. *A Thousand Years of Nonlinear History*. New York: Zone.

deMenocal, Peter B., and Chris Stringer. 2016. "Climate and the Peopling of the World." *Nature*. doi:10.1038/nature19471.

Depew, David J., and Bruce H. Weber. 1997. *Darwinism Evolving: Systems Dynamics and the Genealogy of Natural Selection*. Cambridge, MA: MIT Press.

d'Errico, Francesco, Christopher Henshilwood, Graeme Lawson, Marian Vanhaeren, Anne-Marie Tillier, Marie Soressi, Frédérique Bresson, et al. 2003. "Archaeological Evidence for the Emergence of Language, Symbolism, and Music—an Alternative Multidisciplinary Perspective." *Journal of World Prehistory* 17:1–70.

d'Errico, Francesco, and Marian Vanhaeren. 2007. "Evolution or Revolution? New Evidence for the Origin of Symbolic Behaviour in and out of Africa." In Mellars et al. 2007, 275–86.

Dibble, Harold L., and Ofer Bar-Josef, eds. 1995. *The Definition and Interpretation of Levallois Technology*. Madison, WI: Prehistory Press.

Donald, Merlin. 1991. *Origins of the Modern Mind: Three Stages in the Evolution of Culture and Cognition*. Cambridge, MA: Harvard University Press.

———. 1999. "Preconditions for the Evolution of Protolanguages." In *The Descent of Mind: Psychological Perspectives on Hominid Evolution*, edited by Michael C. Corballis and Stephen E. G. Lea, 138–54. Oxford: Oxford University Press.

Dunbar, Robin. 1998. *Grooming, Gossip, and the Evolution of Language*. Cambridge, MA: Harvard University Press.

Durham, William. 1991. *Coevolution: Genes, Culture, and Human Diversity*. Stanford, CA: Stanford University Press.

El-Hani, Charbel Niño, João Queiroz, and Claus Emmeche. 2009. *Genes, Information, and Semiosis*. Tartu, Estonia: Tartu University Press.

Emmeche, Claus, and Kalevi Kull. 2011. *Towards a Semiotic Biology: Life Is the Action of Signs*. London: Imperial College Press.

Endicott, Phillip, Simon Y. W. Ho, and Chris Stringer. 2010. "Using Genetic Evidence to Evaluate Four Paleoanthropological Hypotheses for the Timing of Neanderthal and Modern Human Origins." *Journal of Human Evolution* 59: 87–95.

Eriksen, Nina, Jacob Tougaard, Lee A. Miller, and David Helweg. 2005. "Cultural

Change in the Songs of Humpback Whales (*Megaptera novaeangliae*) from Tonga." *Behavior* 142:305–25.

Evans, P. D., S. L. Gilbert, N. Mekel-Bobrov, E. J. Vallender, J. R. Anderson, L. M. Vaez-Azizi, S. A. Tishkoff, R. R. Hudson, and B. T. Lahn. 2005. "Microcephalin, a Gene Regulating Brain Size, Continues to Evolve Adaptively in Humans." *Science* 309:1717–20.

Favereau, Donald. 2010. *Essential Readings in Biosemiotics: Anthology and Commentary*. Dordrecht: Springer Verlag.

Finlayson, Clive. 2005. "Biogeography and Evolution of the Genus *Homo*." *Trends in Ecology and Evolution* 8:457–63.

Finlayson, Clive, and José S. Carrión. 2007. "Rapid Ecological Turnover and Its Impact on Neanderthal and Other Human Populations." *Trends in Ecology and Evolution* 22:213–22.

Finn, Julian K., Tom Tregenza, and Mark D. Norman. 2009. "Defensive Tool Use in a Coconut-Carrying Octopus." *Current Biology* 19:R1069–70.

Fodor, Jerry. 1983. *The Modularity of Mind: An Essay on Faculty Psychology*. Cambridge, MA: MIT Press.

———. 1986. "Why Paramecia Don't Have Mental Representations." *Midwest Studies in Philosophy* 10:3–23.

———. 1990. *A Theory of Content and Other Essays*. Cambridge, MA: MIT Press.

———. 2000. *The Mind Doesn't Work That Way: The Scope and Limits of Computational Psychology*. Cambridge, MA: MIT Press.

Forster, Peter. 2004. "Ice Ages and the Mitochondrial DNA Chronology of Human Dispersals: A Review." *Philosophical Transactions of the Royal Society B* 359: 255–64.

Freeberg, Todd M., Andrew P. King, and Meredith J. West. 2001. "Cultural Transmission of Vocal Traditions in Cowbirds (*Molothus ater*) Influences Courtship Patterns and Mate Preferences." *Journal of Comparative Psychology* 115:201–11.

Fu, Qiaomei, Alissa Mittnik, Philip L. F. Johnson, Kirsten Bos, Martina Lari, Ruth Bollongino, Chengkai Sun, et al. 2013. "A Revised Timescale for Human Evolution Based on Ancient Mitochondrial Genomes." *Current Biology* 23: 553–59.

Gamble, Clive. 1999. *The Palaeolithic Societies of Europe*. Cambridge: Cambridge University Press.

———. 2007. *Origins and Revolutions: Human Identity in Earliest Prehistory*. Cambridge: Cambridge University Press.

———. 2012. "When the Words Dry Up: Music and Material Metaphors Half a Million Years Ago." In *Music, Language, and Human Evolution*, edited by Nicholas Bannan, 81–106. Oxford: Oxford University Press.

Gavrilets, Sergey. 2004. *Fitness Landscapes and the Origin of Species*. Princeton, NJ: Princeton University Press.

———. 2010. "High-Dimensional Fitness Landscapes and Speciation." In Pigliucci and Müller 2010, 45–79.

Geertz, Clifford. 1973. *The Interpretation of Cultures*. New York: Basic Books.

Geissmann, Thomas. 2000. "Gibbon Songs and Human Music from an Evolutionary Perspective." In *The Origins of Music*, edited by Nils L. Wallin, Björn Merker, and Steven Brown, 103–23. Cambridge, MA: MIT Press.

Gero, Shane, Hal Whitehead, and Luke Rendell. 2016. "Individual, Unit and Vocal Clan Level Identity Cues in Sperm Whales." *Royal Society Open Science*. doi:10.1098/rsos.150372.

Gibson, Kathleen R. 2007. "Putting It All Together: A Constructionist Approach to the Evolution of Human Mental Capacities." In Mellars et al. 2007, 67–77.

Godfrey-Smith, Peter. 2016. *Other Minds: The Octopus, the Sea, and the Deep Origins of Consciousness*. New York: Farrar, Straus and Giroux.

Gould, Stephen Jay. 1981. *The Mismeasure of Man*. New York: Norton.

———. 1989. *Wonderful Life: The Burgess Shale and the Nature of History*. New York: Norton.

Green, Richard E., Johannes Krause, Adrian W. Briggs, Tomislav Maricic, Udo Stenzel, Martin Kircher, Nick Patterson, et al. 2010. "A Draft Sequence of the Neandertal Genome." *Science* 328:710–23.

Grusin, Richard. 2015. "Radical Mediation." *Critical Inquiry* 42:124–48.

Hammer, Michael F., August E. Woerner, Fernando L. Mendez, Joseph C. Watkins, and Jeffrey D. Wall. 2011. "Genetic Evidence for Archaic Admixture in Africa." *Proceedings of the National Academy of Sciences* 108:15123–28.

Hauser, M. D., N. Chomsky, and W. T. Fitch. 2002. "The Language Faculty: Who Has It, What Is It, and How Did It Evolve?" *Science* 298:1569–79.

Henrich, Joseph, Robert Boyd, and Peter J. Richerson. 2008. "Five Misunderstandings about Cultural Evolution." *Human Nature* 19:119–37.

Henrich, Joseph, and Richard McElreath. 2003. "The Evolution of Cultural Evolution." *Evolutionary Anthropology* 12:123–35.

Henshilwood, Christopher Stuart. 2007. "Fully Symbolic *Sapiens* Behaviour: Innovation in the Middle Stone Age at Blombos Cave, South Africa." In Mellars et al. 2007, 123–32.

Henshilwood, Christopher, Francesco d'Errico, Marian Vanhaeren, Karen van Niekerk, and Zenobia Jacobs. 2004. "Middle Stone Age Shell Beads from South Africa." *Science* 304:404.

Henshilwood, Christopher S., Francesco d'Errico, Karen L. van Niekerk, Yvan Coquinot, Zenobia Jacobs, Stein-Erik Lauritzen, Michel Menu, and Renata García-Moreno. 2011. "A 100,000-Year-Old Ochre-Processing Workshop at Blombos Cave, South Africa." *Science* 334:219–22.

Henshilwood, Christopher S., Francesco d'Errico, Royden Yates, Zenobia Jacobs, Chantal Tribolo, Geoff A. T. Duller, Norbert Mercier, et al. 2002. "Emergence of Modern Human Behavior: Middle Stone Age Engravings from South Africa." *Science* 295:1278–80.

Higham, Thomas, Laura Basell, Roger Jacobi, Rachel Wood, Christopher Bronk Ramsey, and Nicholas J. Conard. 2012. "Testing Models for the Beginnings of the Aurignacian and the Advent of Figurative Art and Music: The Radiocarbon Chronology of Geissenklösterle." *Journal of Human Evolution* 62:664–76.

Hirata, Satoshi, Kunio Watanabe, and Masao Kawai. 2001. "'Sweet-Potato Washing' Revisited." In *Primate Origins of Human Cognition and Behavior*, edited by Tetsuro Matsuzawa, 487–508. Hong Kong: Springer.

Hochberg, Mark S., and Judah Folkman. 1972. "Mechanism of Size Limitation of Bacterial Colonies." *Journal of Infectious Diseases* 126:629–35.

Holekamp, Kay E. 2007. "Questioning the Social Intelligence Hypothesis." *Trends in Cognitive Sciences* 11:65–69.

Hollox, E. J., Mark Poulter, Marek Zvarik, Vladimir Ferak, Amande Krause, Trefor Jenkins, Nilmani Saha, Andrew I. Kozlov, and Dallas M. Swallow. 2001. "Lactase Haplotype Diversity in the Old World." *American Journal of Human Genetics* 68:160–72.

Hsieh, P., A. E. Woerner, J. D. Wall, J. Lachance, S. A. Tishkoff, R. N. Gutenkunst, and M. F. Hammer. 2016. "Model-Based Analyses of Whole-Genome Data Reveal a Complex Evolutionary History Involving Archaic Introgression in Central African Pygmies." *Genome Research* 26:291–300.

Hublin, Jean-Jacques, Abdelouahed Ben-Ncer, Shara E. Bailey, Sarah E. Freidline, Simon Neubauer, Matthew M. Skinner, Inga Bergmann, et al. 2017. "New Fossils from Jebel Irhoud, Morocco and the Pan-African Origin of *Homo sapiens*." *Nature*. doi:10.1038/nature22336.

Hunley, Keith L., Meghan E. Healy, and Jeffrey C. Long. 2009. "The Global Pattern of Gene Identity Variation Reveals a History of Long-Range Migrations, Bottlenecks, and Local Mate Exchange: Implications for Biological Race." *American Journal of Physical Anthropology* 139:35–46.

Huxley, Julian. (1942) 2010. *Evolution: The Modern Synthesis*. Cambridge, MA: MIT Press.

Ingold, Tim. 1993. "The Temporality of the Landscape." *World Archaeology* 25: 152–73.

———. 2000. *The Perception of the Environment: Essays on Livelihood, Dwelling and Skill*. New York: Routledge.

Itan, Yuval, Adam Powell, Mark A. Beaumont, Joachim Burger, and Mark G. Thomas. 2009. "The Origins of Lactase Persistence in Europe." *PLoS Computational Biology*. doi.org/10.1371/journal.pcbi.1000491.

Jackendoff, Ray. 2002. *Foundations of Language: Brain, Meaning, Grammar, Evolution.* Oxford: Oxford University Press.

Jacob, Pierre. 2014. "Intentionality." In *The Stanford Encyclopedia of Philosophy,* Winter 2014 ed., edited by Edward N. Zalta. http://plato.stanford.edu/archives/win2014/entries/intentionality/.

Jacobs, Zenobia, Richard G. Roberts, Rex F. Galbraith, Hilary J. Deacon, Rainer Grün, Alex Mackay, Peter Mitchell, Ralf Vogelsang, and Lyn Wadley. 2008. "Ages for the Middle Stone Age of Southern Africa: Implications for Human Behavior and Dispersal." *Science* 322:733–35.

Jaubert, Jacques, S. Verheyden, D. Genty, M. Soulier, H. Cheng, D. Blamart, C. Burlet, et al. 2016. "Early Neanderthal Constructions Deep in Bruniquel Cave in Southwestern France." *Nature.* doi:10.1038/nature18291.

Johnson, Cara Roure, and Sally McBrearty. 2010. "500,000 Year Old Blades from the Kapthurin Formation, Kenya." *Journal of Human Evolution* 58:193–200.

Jones, Clive G., John H. Lawton, and Moshe Sachak. 1994. "Organisms as Ecosystem Engineers." *Oikos* 69:373–86.

———. 1997. "Positive and Negative Effects of Organisms as Physical Ecosystem Engineers." *Ecology* 78:1946–57.

Kauffman, Stuart A. 1993. *The Origins of Order: Self-Organization and Selection in Evolution.* New York: Oxford University Press.

———. 2000. *Investigations.* Oxford: Oxford University Press.

Kawai, M. 1965. "Newly Acquired Pre-cultural Behavior of the Natural Troop of Japanese Monkeys on Koshima Islet." *Primates* 6:1–30.

Kim, Jaegwon. 1999. "Making Sense of Emergence." *Philosophical Studies* 95:3–16.

Klein, Richard G. 2000. "Archaeology and the Evolution of Human Behavior." *Evolutionary Anthropology* 9:7–36.

———. 2009. *The Human Career: Human Biological and Cultural Origins.* Chicago: University of Chicago Press.

Kockelman, Paul. 2013a. *Agent, Person, Subject, Self: A Theory of Ontology, Interaction, and Infrastructure.* Oxford: Oxford University Press.

———. 2013b. "Information Is the Enclosure of Meaning: Cybernetics, Semiotics, and Alternative Theories of Information." *Language and Communication* 33: 115–27.

———. 2017. "Semiotic Agency." In *Distributed Agency,* edited by N. J. Enfield and Paul Kockelman, 25–38. New York: Oxford University Press.

Kolbert, Elizabeth. 2014. *The Sixth Extinction: An Unnatural History.* New York: Henry Holt.

Kramnick, Jonathan. 2011. "Against Literary Darwinism." *Critical Inquiry* 37: 315–47.

Kroeber, A. L., and Clyde Kluckhohn. 1952. *Culture: A Critical Review of Concepts and Definitions*. Papers of the Peabody Museum of American Archaeology and Ethnology, Harvard University 47, no. 1. Cambridge, MA: Peabody Museum.

Kull, Kalevi. 1999. "Biosemiotics in the Twentieth Century: A View from Biology." *Semiotica* 127:385–414.

———. 2000. "An Introduction to Phytosemiotics: Semiotic Botany and Vegetative Sign Systems." *Sign Systems Studies* 28:326–50.

———. 2009. "Vegetative, Animal, and Cultural Semiosis: The Semiotic Threshold Zones." *Cognitive Semiotics* 4:8–27.

Lahr, Marta Mirazón, and Robert A. Foley. 1998. "Towards a Theory of Modern Human Origins: Geography, Demography, and Diversity in Recent Human Evolution." *Yearbook of Physical Anthropology* 41:137–76.

Laland, Kevin N., Tobias Uller, Marcus W. Feldman, Kim Sterelny, Gerd B. Müller, Armin Moczek, Eva Jablonka, and John Odling-Smee. 2015. "The Extended Evolutionary Synthesis: Its Structure, Assumptions and Predictions." *Proceedings of the Royal Society B*. doi:10.1098/rspb.2015.1019.

Langley, Michelle C., Christopher Clarkson, and Sean Ulm. 2008. "Behavioral Complexity in Eurasian Neanderthal Populations: A Chronological Examination of the Archaeological Evidence." *Cambridge Archaeological Journal* 18:289–307.

Leakey, M. D. 1971. *Olduvai Gorge: Excavations in Beds I and II, 1960–63*. Cambridge: Cambridge University Press.

Leijonhufvud, Axel. 1997. "Models and Theories." *Journal of Economic Methodology* 4:193–98.

Leroi-Gourhan, André. 1993. *Gesture and Speech*. Translated by Anna Bostock Berger. Cambridge, MA: MIT Press.

Lewontin, R. C. 1972. "The Apportionment of Human Diversity." *Evolutionary Biology* 6:381–98.

———. 1983. "Gene, Organism and Environment." In *Evolution from Molecules to Men*, edited by D. S. Bendall, 273–85. Cambridge: Cambridge University Press.

———. 2000. *The Triple Helix: Gene, Organism, and Environment*. Cambridge, MA: Harvard University Press.

Lewontin, R. C., Steven Rose, and Leon J. Kamin. 1984. *Not in Our Genes: Biology, Ideology, and Human Nature*. New York: Pantheon.

Lieberman, Philip. 2006. *Toward an Evolutionary Biology of Language*. Cambridge, MA: Harvard University Press.

Lotman, Juri. 2005. "On the Semiosphere." *Sign Systems Studies* 33:205–29.

MacKay, Donald M. 1969. *Information, Mechanism, and Meaning*. Cambridge, MA: MIT Press.

Malthus, Thomas. (1798) 2008. *An Essay on the Principle of Population*. Edited by Geoffrey Gilbert. Oxford: Oxford University Press.

Maxwell, James Clerk. 1868. "On Governors." *Proceedings of the Royal Society of London* 16:270–83.

Maynard Smith, John. 2000. "The Concept of Information in Biology." *Philosophy of Science* 67:177–94.

Maynard Smith, John, and Eörs Szathmáry. 1995. *The Major Transitions in Evolution*. Oxford: Oxford University Press.

Mayr, Otto. 1971. "Maxwell and the Origins of Cybernetics." *Isis* 62:424–44.

McBrearty, Sally. 2007. "Down with the Revolution!" In Mellars et al. 2007, 133–52.

McBrearty, Sally, and Alison S. Brooks. 2000. "The Revolution That Wasn't: A New Interpretation of the Origin of Modern Human Behavior." *Journal of Human Evolution* 39:453–563.

McGhee, George. 2007. *The Geometry of Evolution: Adaptive Landscapes and Theoretical Morphospaces*. Cambridge: Cambridge University Press.

McNabb, John, Francesca Binyon, and Lee Hazelwood. 2004. "The Large Cutting Tools from the South African Acheulean and the Questions of Social Traditions." *Current Anthropology* 45:653–77.

Mekel-Bobrov, Nitzan, Sandra L. Gilbert, Patrick D. Evans, Eric J. Vallender, Jeffrey R. Anderson, Richard R. Hudson, Sarah A. Tishkoff, and Bruce T. Lahn. 2005. "Ongoing Adaptive Evolution of *ASPM*, a Brain Size Determinant in *Homo sapiens*." *Science* 309:1720–22.

Mekel-Bobrov, N., D. Posthuma, S. L. Gilbert, P. Lind, M. F. Gosso, M. Luciano, S. E. Harris, et al. 2007. "The Ongoing Adaptive Evolution of *ASPM* and Microcephalin Is Not Explained by Increased Intelligence." *Human Molecular Genetics* 16:600–608.

Mellars, Paul. 2006a. "Archaeology and the Dispersal of Modern Humans in Europe: Deconstructing the 'Aurignacian.'" *Evolutionary Anthropology* 15: 167–82.

———. 2006b. "Why Did Modern Human Populations Disperse from Africa *ca*. 60,000 Years Ago? A New Model." *Proceedings of the National Academy of Sciences* 103:9381–86.

———. 2007. "Rethinking the Human Revolution: Eurasian and African Perspectives." In Mellars et al. 2007, 1–11.

Mellars, Paul, Katie Boyle, Ofer Bar-Yosef, and Chris Stringer, eds. 2007. *Rethinking the Human Revolution*. Cambridge: McDonald Institute.

Mesoudi, Alex. 2011. *Cultural Evolution: How Darwinian Theory Can Explain Human Culture and Synthesize the Social Sciences*. Chicago: University of Chicago Press.

Millikan, Ruth Garrett. 1989. "Biosemantics." *Journal of Philosophy* 86:281–97.

Mitchell, Sandra D. 2009. *Unsimple Truths: Science, Complexity, and Policy*. Chicago: University of Chicago Press.

Mithen, Steven. 1996. *The Prehistory of the Mind: The Cognitive Origins of Art and Science*. London: Thames and Hudson.

Morley, Iain. 2009. "Rituals and Music: Parallels and Practice, and the Palaeolithic." In *Becoming Human: Innovation in Prehistoric Material and Spiritual Culture*, edited by Colin Renfrew and Iain Morley, 159–78. Cambridge: Cambridge University Press.

———. 2013. *The Prehistory of Music: Human Evolution, Archaeology, and the Origins of Musicality*. Oxford: Oxford University Press.

Müller, Ulrich C., Jörg Pross, Polychronis C. Tzedakis, Clive Gamble, Ulrich Kotthoff, Gerhard Schmiedl, Sabine Wulf, and Kimon Christanis. 2011. "The Role of Climate in the Spread of Modern Humans into Europe." *Quaternary Science Reviews* 30:273–79.

Nagel, Thomas. 1974. "What Is It Like to Be a Bat?" *Philosophical Review* 83:435–50.

Nettl, Bruno. 2000. "An Ethnomusicologist Contemplates Universals in Musical Sound and Musical Culture." In *The Origins of Music*, edited by Nils L. Wallin, Björn Merker, and Steven Brown, 463–72. Cambridge, MA: MIT Press.

Nielsen, Rasmus. 2009. "Adaptationism—30 Years after Gould and Lewontin." *Evolution* 63:2487–90.

Nitecki, Matthew H., and Doris Nitecki, eds. 1992. *History and Evolution*. Albany: SUNY Press.

Nöth, Winfried. 2011. "From Representation to Thirdness and Representamen to Medium: Evolution of Peircean Key Terms and Topics." *Transactions of the Charles S. Peirce Society* 47:445–81.

Nowell, April. 2010. "Defining Behavioral Modernity in the Context of Neandertal and Anatomically Modern Human Populations." *Annual Review of Anthropology* 39:437–52.

O'Connell, James F., and Jim Allen. 2007. "Pre-LGM Sahul (Pleistocene Australia–New Guinea) and the Archaeology of Early Modern Humans." In Mellars et al. 2007, 395–410.

Odling-Smee, F. John, Kevin N. Laland, and Marcus W. Feldman. 2003. *Niche Construction: The Neglected Process in Evolution*. Princeton, NJ: Princeton University Press.

Okanoya, Kazuo. 2002. "Sexual Display as a Syntactical Vehicle: The Evolution of Syntax in Birdsong and Human Language through Sexual Selection." In *The Transition to Language*, edited by Alison Wray, 46–63. Oxford: Oxford University Press.

Okanoya, K., and A. Yamaguchi. 1997. "Adult Bengalese Finches (*Lonchura striata* var. *domestica*) Require Real-Time Auditory Feedback to Produce Normal Song Syntax." *Journal of Neurobiology* 33:343–56.

Ouattara, Karim, Alban Lemasson, and Klaus Zuberbühler. 2009. "Campbell's

Monkeys Concatenate Vocalizations into Context-Specific Call Sequences." *Proceedings of the National Academy of Sciences USA*. doi:10.1073/pnas.0908118106.

Pavličev, Mihaela, Richard O. Prum, Gary Tomlinson, and Günter Wagner. 2016. "Systems Emergence: The Origin of Individuals in Biological and Biocultural Evolution." In *Evolutionary Theory: A Hierarchical Perspective*, edited by Niles Eldredge, Telmo Pievani, Emanuele Serrelli, and Ilya Temkin, 202–23. Chicago: University of Chicago Press.

Payne, Katherine, Peter Tyack, and Roger Payne. 1983. "Progressive Changes in the Songs of Humpback Whales: A Detailed Analysis of Two Seasons in Hawaii." In *Communication and Behavior in Whales*, edited by Roger Payne, 9–57. Boulder, CO: Westview Press.

Peirce, Charles S. 1955. *Philosophical Writings of Peirce*. Edited by Justus Buchler. New York: Dover.

———. 1958. *Selected Writings: Values in a Universe of Chance*. Edited by Philip P. Wiener. New York: Dover.

———. 1998. *The Essential Peirce: Selected Philosophical Writings*. Edited by the Peirce Edition Project. 2 vols. Bloomington: Indiana University Press.

Perreault, Charles. 2012. "The Pace of Cultural Evolution." *PLoS ONE*. doi:org/10.1371/journal.pone.0045150.

Perry, Clint J., Andrew B. Barron, and Ken Cheng. 2013. "Invertebrate Learning and Cognition: Relating Phenomena to Neural Substrate." *Cognitive Science*. doi:10.1002/wcs.1248.

Pigliucci, Massimo, and Jonathan Kaplan. 2006. *Making Sense of Evolution: The Conceptual Foundations of Evolutionary Biology*. Chicago: University of Chicago Press.

Pigliucci, Massimo, and Gerd B. Müller, eds. 2010. *Evolution: The Extended Synthesis*. Cambridge, MA: MIT Press.

Pinker, Steven. 1994. *The Language Instinct: How the Mind Creates Language*. New York: Harper.

———. 1997. *How the Mind Works*. New York: Norton.

Pollard, Thomas D., and William C. Earnshaw. 2008. *Cell Biology*. 2nd ed. Philadelphia: Saunders/Elsevier

Pope, Matt, and Mark Roberts. 2005. "Observations on the Relationship between Palaeolithic Individuals and Artefact Scatters at the Pleistocene Site of Boxgrove, UK." In *The Hominid Individual in Context: Archaeological Investigations of Lower and Middle Palaeolithic Landscapes, Locales, and Artefacts*, edited by Clive Gamble and Martin Porr, 81–97. New York: Routledge.

Posth, Cosimo, Christoph Wißing, Keito Kitagawa, Luca Pagani, Laura van Holstein, Fernando Racimo, Kurt Wehrberger, et al. 2017. "Deeply Divergent

Archaic Mitochondrial Genome Provides Lower Time Boundary for African Gene Flow into Neanderthals." *Nature Communications*. doi:10.1038/ncomms16046.

Potts, Richard. 2013. "Hominin Evolution in Settings of Strong Environmental Variability." *Quaternary Science Reviews* 73:1–13.

Powell, Adam, Stephen Shennan, and Mark G. Thomas. 2009. "Late Pleistocene Demography and the Appearance of Modern Human Behavior." *Science* 324: 1298–1301.

Price, Douglas T., and Ofer Bar-Yosef, eds. 2011. "The Origins of Agriculture: New Data, New Ideas." Special issue, *Current Anthropology* 52, supplement 4.

Prum, Richard O. 2017. *The Evolution of Beauty: How Darwin's Forgotten Theory of Mate Choice Shapes the Animal World—and Us*. New York: Norton.

Reich, David, Richard E. Green, Martin Kircher, Johannes Krause, Nick Patterson, Eric Y. Durand, Bence Viola, et al. 2010. "Genetic History of an Archaic Hominin Group from Denisova Cave in Siberia." *Nature* 468:1053–60.

Rendell, Luke, Laurel Fogarty, and Kevin N. Laland. 2011. "Runaway Cultural Niche Construction." *Philosophical Transactions of the Royal Society B* 366: 823–35.

Renfrew, Colin. 2008. "Neuroscience, Evolution, and the Sapient Paradox." *Philosophical Transactions of the Royal Society B* 363:2041–47.

Renfrew, Colin, Chris Firth, and Lambros Malafouris, eds. 2008. "The Sapient Mind: Archaeology Meets Neuroscience." Special issue, *Philosophical Transactions of the Royal Society B* 363.

Richardson, Sarah S. 2011. "Race and IQ in the Postgenomic Age: The Microcephaly Case." *BioSocieties* 6:420–46.

Richerson, Peter J., and Robert Boyd. 2005. *Not by Genes Alone: How Culture Transformed Human Evolution*. Chicago: University of Chicago Press.

———. 2013. "Rethinking Paleoanthropology: A World Queerer than We Supposed?" In *Evolution of Mind, Brain, and Culture*, edited by Gary Hatfield and Holly Pittman, 263–302. Philadelphia: University of Pennsylvania Museum of Archaeology and Anthropology.

Richerson, Peter J., Robert Boyd, and Robert L. Bettinger. 2009. "Cultural Innovations and Demographic Change." *Human Biology* 81:211–35.

Richter, Daniel, Rainer Grün, Renaud Joannes-Boyau, Teresa E. Steele, Fehti Amani, Mathieu Rué, Paul Fernandes, et al. 2017. "The Age of the Hominin Fossils from Jebel Irhoud, Morocco, and the Origins of the Middle Stone Age." *Nature*. doi:10.1038/nature22335.

Robertson, Douglas S. 1991. "Feedback Theory and Darwinian Evolution." *Journal of Theoretical Biology* 152:469–84.

Robertson, Douglas S., and Michael C. Grant. 1996. "Feedback and Chaos in Darwinian Evolution." *Complexity* 2, no. 1: 10–14; 2, no. 2: 18–30.

Rodríguez-Vidal, Joaquín, Francesco d'Errico, Francisco Giles Pacheco, Ruth Blasco, Jordi Rosell, Richard P. Jennings, Alain Queffelec, et al. 2014. "A Rock Engraving Made by Neanderthals in Gibraltar." *Proceedings of the National Academy of Sciences* 111:13301–6.

Rodseth, L., R. W. Wrangham, A. Harrigan, and B. B. Smuts. 1991. "The Human Community as a Primate Society." *Current Anthropology* 32:221–54.

Roebroeks, Wil, Mark J. Sier, Trine Kellberg Nielsen, Dimitri De Loecker, Josep Maria Pares, Charles E. S. Arps, and Herman J Mücher. 2012. "Use of Red Ochre by Early Neandertals." *Proceedings of the National Academy of Sciences* 109:1889–94.

Rumbaugh, Duane M., James E. King, Michael J. Beran, David A. Washburn, and Kristy L. Gould. 2007. "A Salience Theory of Learning and Behavior: With Perspectives on Neurobiology and Cognition." *International Journal of Primatology* 28: 973–96.

Sahlins, Marshall. 1977. *The Use and Abuse of Biology: An Anthropological Critique of Sociobiology*. Ann Arbor: University of Michigan Press.

Schiffels, S., and R. Durbin. 2014. "Inferring Human Population Size and Separation History from Multiple Genome Sequences." *Nature Genetics* 46: 919–25.

Schrödinger, Erwin. (1944) 1992. *What Is Life? With "Mind and Matter" and Autobiographical Sketches*. Cambridge: Cambridge University Press.

Seyfarth, R. M., D. L. Cheyney, and P. Marler. 1980. "Monkey Responses to Three Different Alarm Calls: Evidence of Predator Classification and Semantic Communication." *Science* 210:801–3.

Shannon, Claude E., and Warren Weaver. 1949. *The Mathematical Theory of Communication*. Urbana: University of Illinois Press.

Shea, John. 2007. "The Boulevard of Broken Dreams: Evolutionary Discontinuity in the Late Pleistocene Levant." In Mellars et al. 2007, 219–32.

———. 2011. "*Homo sapiens* Is as *Homo sapiens* Was: Behavioral Variability versus 'Behavioral Modernity' in Paleolithic Archaeology." *Current Anthropology* 52:1–35.

Shennan, Stephen. 2001. "Demography and Cultural Innovation: A Model and Its Implications for the Emergence of Modern Human Culture." *Cambridge Archaeological Journal* 11:5–16.

———. 2002. *Genes, Memes and Human History: Darwinian Archaeology and Cultural Evolution*. London: Thames and Hudson.

Short, T. L. 2007. *Peirce's Theory of Signs*. Cambridge: Cambridge University Press.

Shryock, Andrew, and Daniel Lord Smail. 2011. *Deep History: The Architecture of Past and Present*. Berkeley: University of California Press.

Shumaker, Robert W., Kristina R. Walkup, and Benjamin B. Beck. 2011. *Animal*

Tool Behavior: The Use and Manufacture of Tools by Animals. Baltimore: Johns Hopkins University Press.
Silverstein, Michael. 1993. "Metapragmatic Discourse and Metapragmatic Function." In *Reflexive Language: Reported Speech and Metapragmatics*, edited by John A. Lucy, 33–58. Cambridge: Cambridge University Press.
———. 2003. "Indexical Order and the Dialectics of Sociolinguistic Life." *Language and Communication* 23:193–229.
Smail, Daniel Lord. 2008. *On Deep History and the Brain*. Berkeley: University of California Press.
Smolker, Rachel, Andrew Richards, Richard Connor, Janet Mann, and Per Berggren. 1997. "Sponge Carrying by Dolphins (Delphinidae, *Tursiops* sp.): A Foraging Specialization Involving Tool Use?" *Ethology* 103:454–65.
Sperber, Dan. 1996. *Explaining Culture: A Naturalistic Approach*. Oxford: Blackwell.
Steele, James, and Stephen Shennan. 2009. "Introduction: Demography and Cultural Macroevolution." *Human Biology* 81:105–19.
Sterelny, Kim. 2000. "The 'Genetic Program' Program: A Commentary on Maynard Smith on Information in Biology." *Philosophy of Science* 67:195–201.
———. 2012. *The Evolved Apprentice: How Evolution Made Humans Unique*. Cambridge, MA: MIT Press.
Stone, Linda, and Paul F. Lurquin. 2007. *Genes, Culture, and Human Evolution: A Synthesis*. Oxford: Oxford University Press.
Stringer, Chris. 2007. "The Origin and Dispersal of *Homo sapiens*: Our Current State of Knowledge." In Mellars et al. 2007, 15–20.
Tambiah, Stanley J. 1979. "A Performative Approach to Ritual." *Proceedings of the British Academy* 65:113–69.
Texier, J.-P., Guillaume Porraz, John Parkington, Jean-Philippe Rigaud, Cedric Poggenpoel, Christopher Miller, Chantal Tribolo, et al. 2010. "A Howiesons Poort Tradition of Engraving Ostrich Eggshell Containers Dated to 60,000 Years Ago at Diepkloof Rock Shelter, South Africa." *Proceedings of the National Academy of Sciences* 107:6180–85.
Thompson, Evan. 2007. *Mind in Life*. Cambridge, MA: Harvard University Press.
Timmermann, Axel, and Tobias Friedrich. 2016. "Late Pleistocene Climate Drivers of Early Human Migration." *Nature*. doi:10.1038/nature19365.
Tishkoff, S. A., R. Varkonyi, N. Cahinhinan, S. Abbes, G. Argyropoulos, G. Destro-Bisol, A. Drousiotou, et al. 2001. "Haplotype Diversity and Linkage Disequilibrium at Human *G6PD*: Recent Origin of Alleles That Confer Malarial Resistance." *Science* 293:455–62.
Tomasello, Michael. 2008. *Origins of Human Communication*. Cambridge, MA: MIT Press.
Tomasello, M., M. Carpenter, J. Call, T. Behne, and H. Moll. 2005. "Understanding

and Sharing Intentions: The Origins of Cultural Cognition." *Behavioral and Brain Sciences* 28:675–735.

Tomlinson, Gary. 2015. *A Million Years of Music: The Emergence of Human Modernity*. New York: Zone.

———. 2017. "Two Deep-Historical Models of Climate Crisis." *South Atlantic Quarterly* 116:19–31.

Tremlin, Todd. 2006. *Minds and Gods: The Cognitive Foundations of Religion*. Oxford: Oxford University Press.

Vanhaeren, Marian, and Francesco d'Errico. 2006. "Aurignacian Ethno-linguistic Geography of Europe Revealed by Personal Ornaments." *Journal of Archaeological Science* 33:1105–28.

Van Valen, Leigh. 1973. "A New Evolutionary Law." https://dl.dropboxusercontent.com/u/18310184/about-leigh-van-valen/Piglet%20Papers/1973%20new%20evol%20law.pdf.

Von Baeyer, Hans Christian. 2003. *Information: The New Language of Science*. Cambridge, MA: Harvard University Press.

Wade, Nicholas. 2014. *A Troublesome Inheritance: Genes, Race, and Human History*. New York: Penguin.

Wagner, Günter P. 2014. *Homology, Genes, and Evolutionary Innovation*. Princeton, NJ: Princeton University Press.

Wallace, Alfred Russel. 1858. "On the Tendency of Varieties to Depart Indefinitely from the Original Type." Edited by Charles H. Smith. http://people.wku.edu/charles.smith/wallace/S043.htm.

Wang, S., J. Lachance, S. A. Tishkoff, J. Hey, and J. Xing. 2013. "Apparent Variation in Neanderthal Admixture among African Populations Is Consistent with Gene Flow from Non-African Populations." *Genome Biology and Evolution* 5:2075–81.

White, Mark, and Nick Ashton. 2003. "Lower Palaeolithic Core Technology and the Origins of the Levallois Method in North-Western Europe." *Current Anthropology* 44:598–609.

Wiener, Norbert. 1948. *Cybernetics; or, Control and Communication in the Animal and the Machine*. Cambridge, MA: MIT Press.

Wilde, Sandra, Adrian Timpson, Karola Kirsanow, Elke Kaiser, Manfred Kayser, Martina Unterländer, Nina Hollfelder, et al. 2014. "Direct Evidence for Positive Selection of Skin, Hair, and Eye Pigmentation in Europeans during the Last 5,000y." *Proceedings of the National Academy of Sciences* 111:4832–37.

Wilkinson, David M. 2006. *Fundamental Processes in Ecology: An Earth Systems Approach*. Oxford: Oxford University Press.

Williams, Heather, Iris I. Levin, D. Ryan Norris, Amy E. M. Newman, and Nathaniel T. Wheelwright. 2013. "Three Decades of Cultural Evolution in Savannah Sparrow Songs." *Animal Behavior* 85:213–23.

Wilson, Edward O. (1975) 2000. *Sociobiology: The New Synthesis*. Cambridge, MA: Harvard University Press.

———. 1978. *On Human Nature*. Cambridge, MA: Harvard University Press.

Woods, R. P., N. B. Freimer, J. A. De Young, S. C. Fears, N. L. Sicotte, S. K. Service, D. J. Valentino, A. W. Toga, and J. C. Mazziotta. 2006. "Normal Variants of Microcephalin and *ASPM* Do Not Account for Brain Size Variability." *Human Molecular Genetics* 15:2025–29.

Wrangham, Richard. 2009. *Catching Fire: How Cooking Made Us Human*. New York: Basic Books.

Wynn, Thomas. 2002. "Archaeology and Cognitive Evolution." *Behavior and Brain Sciences* 25:389–438.

Wynn, Thomas, and Frederick L. Coolidge. 2004. "The Expert Neandertal Mind." *Journal of Human Evolution* 46:467–87.

Zhang, Juzhong, Garman Harbottle, Changsui Wang, and Zhaochen Kong. 1999. "Oldest Playable Musical Instruments Found at Jiahu Early Neolithic Site in China." *Nature* 401:366–68.

Zubrow, E. 1989. "The Demographic Modeling of Neanderthal Extinction." In *The Human Revolution: Behavioral and Biological Perspectives on the Origins of Modern Humans*, edited by P. Mellars and C. Stringer, 212–31. Princeton, NJ: Princeton University Press.

INDEX